Science in a Nanosecond

Science in a Nanosecond

Illustrated Answers to 100 Basic Science Questions

James A. Haught
Illustrated by Gary T. Hahn

Prometheus Books
Buffalo, New York

SCIENCE IN A NANOSECOND: ILLUSTRATED ANSWERS TO 100 BASIC SCIENCE QUESTIONS. Copyright © 1990 by James A. Haught. All rights reserved. No part of this book may be reproduced in any manner whatsoever without written permission, except in the case of brief quotations embodied in critical articles and reviews. Inquiries should be addressed to Prometheus Books, 700 E. Amherst Street, Buffalo, New York 14215, 716-837-2475.

Library of Congress Cataloging-in-Publication Data

Haught, James A.
   Science in a nanosecond : illustrated answers to 100 basic science questions / James A. Haught ; illustrated by Gary Hahn.
     p.   cm.
   Summary: Uses a question and answer format, with illustrations, to explain why the sky is blue, what a rainbow is, what atoms are, how gravity works, and many other scientific facts and events.

   1. Science—Miscellanea—Juvenile literature. [1. Science—Miscellanea. 2. Questions and answers.] I. Hahn, Gary, ill. II. Title.
Q163.H29   1990
500—dc20
          ISBN 978-0-87975-637-6

90-44186
CIP
AC

For Joel, Jake, Jeb, and Cass—
and their mother, Nancy

Introduction

The syrupy song goes: "Tell me why the stars do shine. Tell me why the ivy twines. Tell me why the skies are blue. And I will tell why I love you."

In response, a college student wrote: "Nuclear fusion makes the stars to shine. Tropisms make the ivy twine. Rayleigh scattering makes the skies so blue. And glandular hormones are why I love you."

Science awareness of this sort (with or without the levity) is needed by everyone to understand our world and the amazing technological revolutions that are changing our lives.

No one is truly educated without a grasp of science fundamentals. They have deep philosophical implications and open the mind to new ways of understanding. For example, it is *awesome* to realize that—

- "Solid" matter is so empty that Planet Earth could be compressed to the size of a *pearl*, if it were squeezed to the threshold at which it began to become a black hole.
- An explosive lightning storm is the unleashed power of detached electrons—yet these same electrons are peaceful in every atom of your body.
- "Maglev" trains are held in midair by a different power of electrons: their synchronized spin in magnets.
- An amount of matter smaller than a *dime* turned into the energy that killed 140,000 people at Hiroshima in 1945.
- As you sit "still," you are moving 550,000 miles per hour with the Milky Way's rotation and 1.3 million miles per hour with the galaxy's movement through the universe.
- Your body contains about *ten billion miles* of DNA, the

threadlike molecules that bear your genetic code.

Science-minded people perceive "the poetry by which they are surrounded," philosopher Herbert Spencer wrote. They look at a drop of water and know "that its elements are held together by a force which, if suddenly liberated, would produce a flash of lightning." When they see a rock with parallel scratches, he added, they know "that over this rock a glacier slid a million years ago."

Science doesn't only give a new vision of reality—it also transforms society. Consider this comparison:

There was no electricity in the Appalachian farm town where I was born in the 1930s, and only a lucky few had gas. Families lived with outdoor toilets, kerosene lamps, handle pumps, wood stoves, and chamber pots. Horses and wagons crunched over unpaved streets. Nobody had heard of television, jet planes, plastic, or moon landings. Today I work on a video terminal in a newspaper office, with a roof dish filling my screen with news from a fixed-orbit satellite 22,300 miles above the equator. The news reports tell of genetic engineering, fusion power, and quasar sightings. Conditions have advanced more during my lifetime than in many preceding centuries.

The change from horse-drawn times to today's world of space travel and 100-channel television sets was purely a result of science. And the changes will accelerate in the 1990s and the 2000s. Science knowledge will be increasingly vital to national progress and to more people's careers.

But trouble looms. Young Americans are falling behind students of other nations in science and math abilities. International tests repeatedly find Americans near the bottom in subjects crucial to high-tech life. Present methods of teaching aren't good enough.

The purpose of this book is to make science fundamentals easy to grasp. When my four children were small, I made simple diagrams so they could see scientific explanations immediately, without wading through mind-boggling chapters. Pictures teach in a nanosecond.

Even if you aren't planning a technical career, you need the wisdom of science. Young people search for meaning in life, and science is the best route. It is the most honest field. Politics, religion, and the arts are realms of opinion—but science deals in facts that are tested and retested until confirmed, then put to work for humanity. Past discoveries become the basis of new ones. Dr. Laurence Peter, author of *The Peter Principle*, said: "The history of science is the only history which displays cumulative progress of knowledge, hence the progress of science is the only yardstick by which we can measure the progress of mankind."

The value of discoveries often isn't known until later. When the first balloon launch occurred in Paris in 1783, scoffing bystanders asked what good it was. A visitor in the city, Benjamin Franklin, replied: "What good is a newborn baby?"

To Europeans in the early 1600s, the foremost event was the Thirty Years War in which Protestants and Catholics slaughtered each other by the millions. Today, few remember that war—but the whole world knows a different legacy from the early 1600s: Galileo's truth about falling bodies and planetary orbits. Similarly, physicist Richard Feynman said that a few hundred years from now, "the most significant event of the 19th century will be judged as Maxwell's discovery of the laws of electrodynamics. The American Civil War will pale into provincial insignificance in comparison."

The average human lifespan was only forty-eight years at the start of this century. Now, thanks to science, it is nearing eighty—and it may pass one hundred in the coming century. People in the past, and in prehistoric times, had the same brain development as modern humans, but they couldn't reap the benefit of snowballing advances. Science writer Arthur C. Clarke (the first person to envision fixed-orbit communications satellites) wrote:

> It is one of the strangest of all facts, impossible for the sensitive mind to contemplate without melancholy, that for at least 50,000 years there have been men on this planet who could conduct a symphony orchestra, discover theorems in pure mathematics, act as secretaries of the United Nations, or pilot a spaceship—had they been given a chance.

Since the dawn of time, animals with eyes have lived by light. But it was only a heartbeat ago that people learned that visible light is just a tiny band in the electromagnetic spectrum. Longer wavelengths make radio waves, microwaves, or infrared rays; shorter ones make ultraviolet rays, X-rays, or gamma rays. Engineer-philosopher R. Buckminster Fuller commented:

> Up to the 20th century, "reality" was everything humans could touch, smell, see and hear. Since the initial publication of the chart of the electromagnetic spectrum, humans have learned that what they can touch, smell, see and hear is less than one-

millionth of reality. Ninety-nine percent of all that is going to affect our tomorrows is being developed by humans using instruments and working in ranges of reality that are non-humanly sensible.

Those who know science perceive such realities, and they share the adventure of the onrushing future. Scholar Jacob Bronowski wrote, "We are a scientific civilization. That means a civilization in which knowledge and its integrity are crucial. Science is only a Latin word for knowledge. . . . Knowledge is our destiny."

Q: What makes summer—

—and winter?

A: The 23.5-degree tilt of Earth's axis.

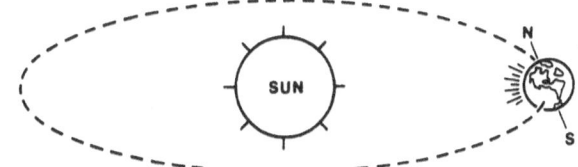

When Earth is on one side of its orbit, the northern hemisphere is tipped toward the sun, so radiation hits more squarely, during about 15 hours of daylight per day—causing summer in the northern hemisphere and winter in the southern hemisphere.

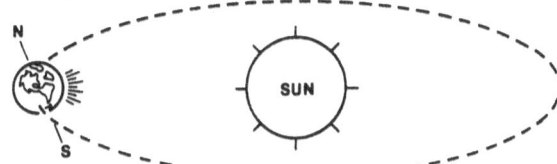

On the opposite side of the orbit, when the northern hemisphere is tipped away from the sun, radiation hits at a low angle with less warming power, for only about 9 hours per day—producing winter in the north and summer in the south.

Q: Since space is inky black—

—what makes the sky blue in the daytime?

A: A principle called "scattering."

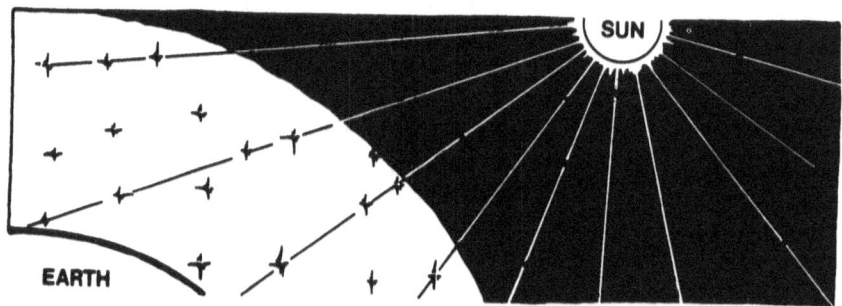

English physicist John Strutt, Baron Rayleigh (1842-1919), discovered that sunlight striking molecules in the atmosphere deflects a high-frequency portion of the light—blue—down to Earth's surface.

(That's why sunsets are red. When the sun is low, its rays pass through more air, shedding more blue, until mostly low-frequency red remains.)

Q: Amid the multitude of stars—

—how can you find the planets?

A: Search along the path followed by the moon.

The solar system is relatively flat. It occupies a line in the sky, called the *ecliptic*, and an accompanying band, the *zodiac*. Planets roam this strip.

Q: When you see "the old moon in the new moon's arms"—

—what's the dim light on the back of the moon?

A: It's "earthshine" reflected from the gleaming side of Earth you can't see.

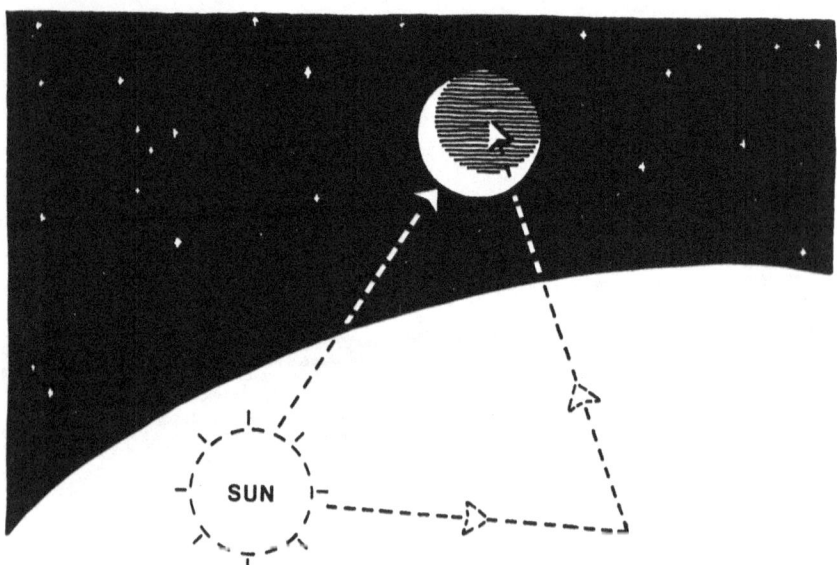

Grasping this, while gazing at the twilight sky, imparts a wondrous three-dimensional sense of the solar system.

Q: How can you use the night sky as a compass?

A: Learn to find the North Star (Polaris).

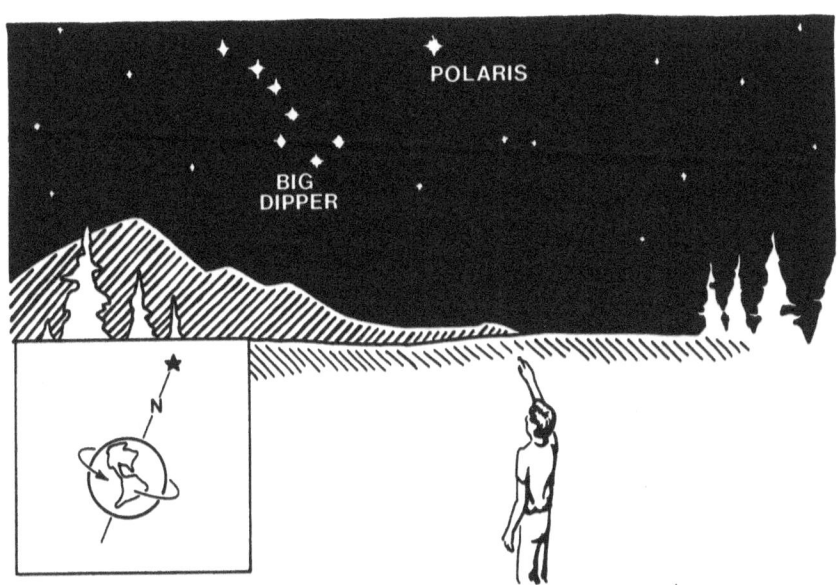

The "pointer stars" at the end of the Big Dipper's cup aim *almost* at Polaris. It's the only star that never seems to move, because it is in line with Earth's rotational axis—thus it is squarely over the North Pole.

Q: What is the Milky Way?

A: This faint-lit band containing billions of stars is actually a view through the dense plane of the spiral galaxy that contains our solar system.

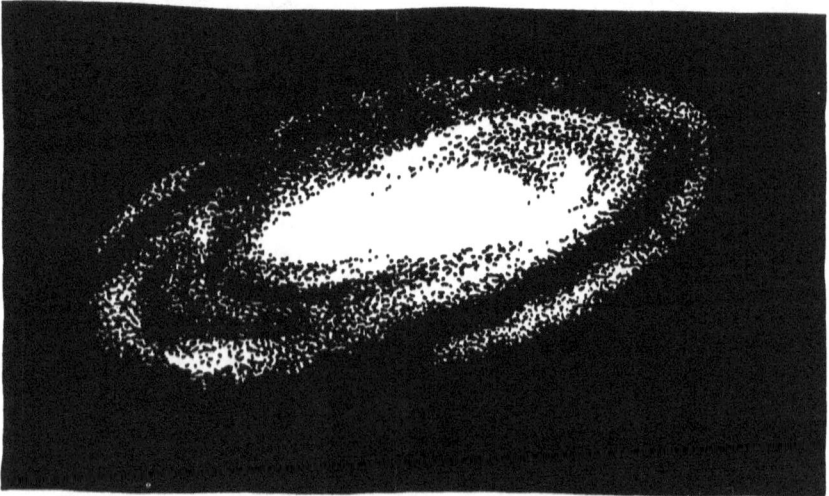

Beyond our galaxy are at least 10 billion other colossal pinwheels and clusters, at mind-boggling distances.

Q: When you're lying "still," how fast are you moving?

A: At the equator you're traveling 1,000 miles per hour with Earth's daily rotation (less elsewhere)

—and 67,000 miles per hour with Earth's yearly orbit of the sun
—and 550,000 miles per hour with the solar system's journey around the Milky Way galaxy
—and 1.3 million miles per hour with the galaxy's motion through the universe.

In contrast, a bullet travels only about 3,000 miles per hour.

Q: With your eyes—

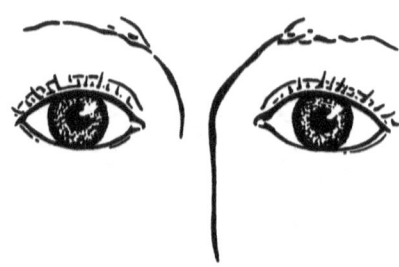

—can you look into the past?

A: Yes—because space distances are so vast, and it takes light so long to reach Earth at its speed of 186,282 miles per second—you see stars as they were centuries ago.

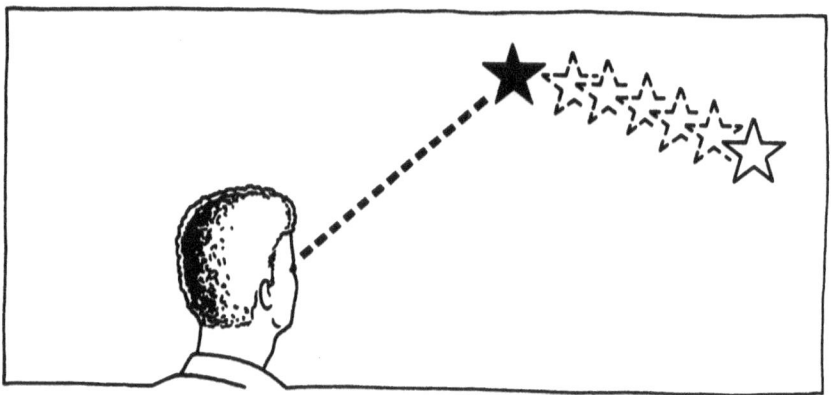

For instance, Polaris is 680 light-years away, so you're seeing light that left Polaris in the 1300s, medieval times, and the star has moved since then.

In 1987, astronomers saw a supernova explode in the Magellanic Clouds outside the Milky Way—but it actually exploded 163,000 years ago, before the first Neanderthal Man.

Q: Anywhere you go on Earth—

—what is the inescapable force?

A: Gravity—the same force that shapes galaxies and solar systems—penetrates every locked room and pulls on every molecule.

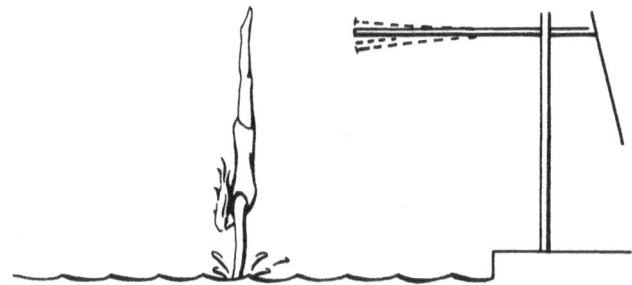

It holds us in our beds, on our chairs, and finally in our graves.

It causes a river to run 1,000 miles sidewise to reach a level a few feet lower.

Q: Which falls faster, a 100-pound cannonball or a 1-ounce bullet?

A: They fall at exactly the same speed.

When Galileo demonstrated this in the 1500s (at the Leaning Tower of Pisa, according to legend), holy scholars refused to believe it and accused him of promulgating evil.

Galileo later found that objects fall ever-faster at the rate of 32 feet per second per second—16 feet the first second, 48 the next, 80 the third, 112 the fourth. (That's why a fall off a chair is minor, but a fall off a skyscraper isn't.)

Q: If a bullet is dropped, and another is shot horizontally—

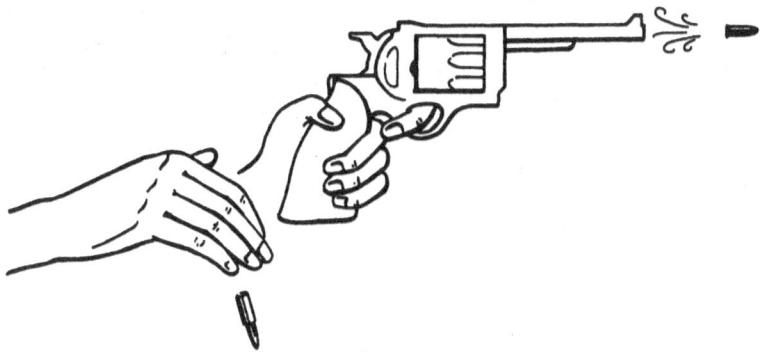

—which reaches the ground first?

A: They land at the same time, because the pull of gravity is the same on each.

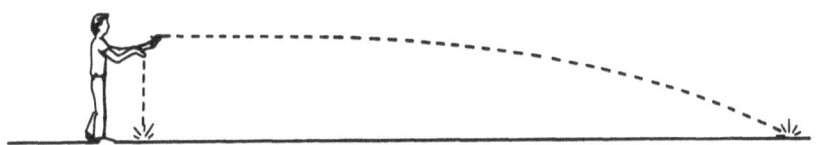

Both fall at 32 feet per second per second.

Q: How does gravity form solar systems—

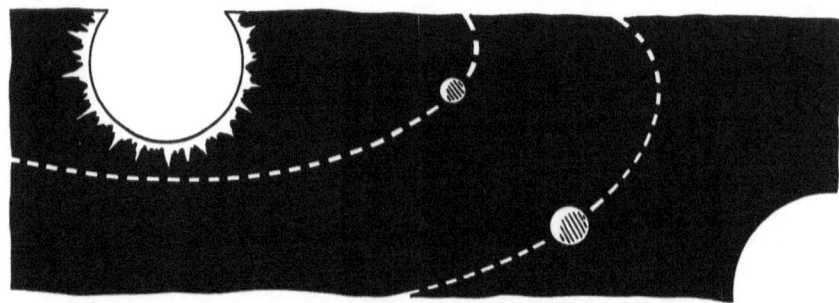

—and galaxies?

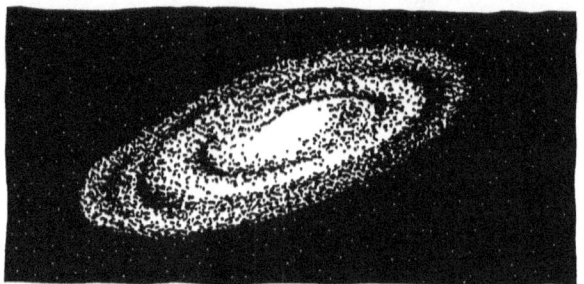

A: By holding bodies in orbits.

Isaac Newton, in a historic flash of genius, saw that orbiting objects "fall." He observed a cannonball's trajectory—

—and realized that, if a cannon shot far enough into the sky, the pull of gravity would keep the ball "falling" forever around Earth. An orbit is a perfectly balanced tug-of-war between inertia, which keeps objects moving in a straight line, and gravity, which draws them toward a massive center.

Q: What gravity-created orbit makes instant communications possible around the globe?

A: Fixed-orbit satellites are reflector stations bouncing millions of television, radio, and telephone signals.

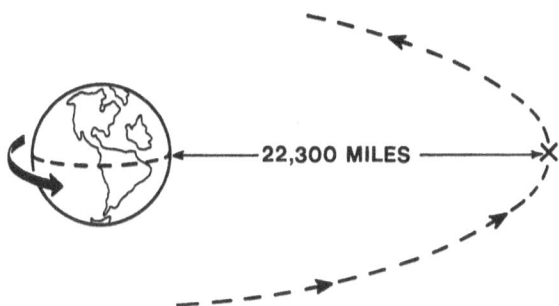

In the 1940s, science writer Arthur C. Clarke was first to perceive that a satellite 22,300 miles above the equator would move exactly in time with Earth's rotation, thus it would "hold still" in the sky and serve as a perfect relay post.

The first such *geosynchronous* satellite was put into orbit in 1964. Now there are hundreds—some relaying up to 120,000 telephone calls simultaneously.

Q: As you go about daily life—how much weight is pressing on you?

A: More than 10 tons.

We live at the bottom of "an ocean of air." Gravity's pull on the air molecules creates a pressure of 14.7 pounds per square inch at sea level. This pushes on every square inch of our bodies, from all directions.

But the ten-ton total squeeze doesn't crush us, because equal pressure inside every organ and cell pushes back. (That's why a person will explode if exposed to the vacuum of space.)

Q: What happens when air is pumped out of a container, removing the internal pressure?

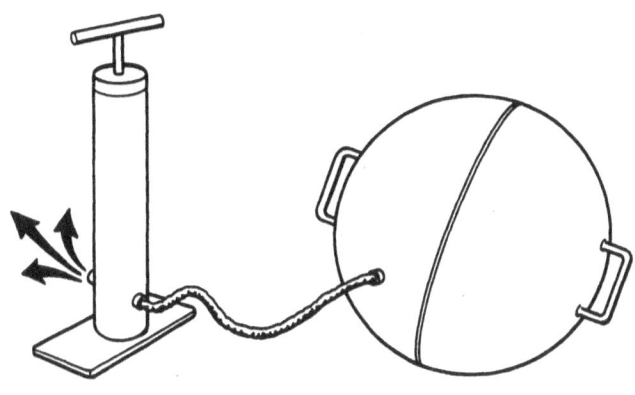

A: The surrounding pressure clamps an awesome vise on the container—about one ton of squeeze on a one-foot ball.

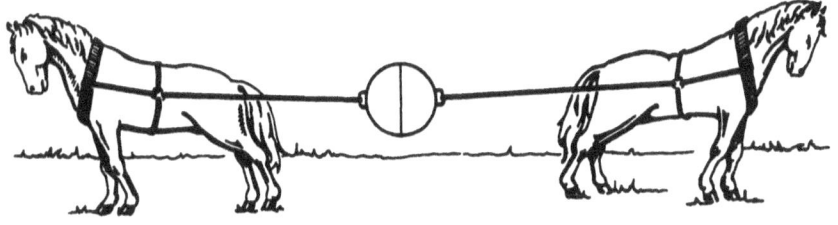

In a famous experiment in 1654 at Magdeburg, Germany, Otto von Guericke pumped air from two snug-fitting hemispheres—and horses couldn't pull them apart. Yet, when a valve was opened and air was allowed inside, the halves fell apart easily.

Q: What "scales" weigh the air—and also predict weather?

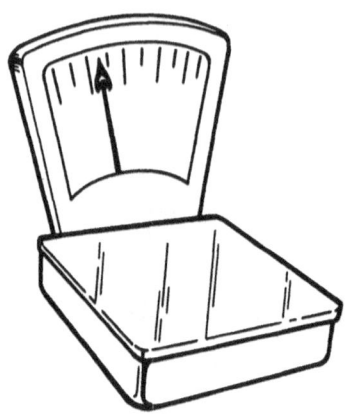

A: Barometers.

Italian physicist Evangelista Torricelli discovered in 1643 that mercury in a long test-tube inverted in a cup of mercury will run down until the air pressure balances it.

At sea level, the column holds at an average of 29.9 inches. On an 8,000-foot mountain, it's only 22.3 inches.

When low atmospheric pressure causes the mercury to sink, stormy weather is approaching. Conversely, a "rising glass" means fair weather.

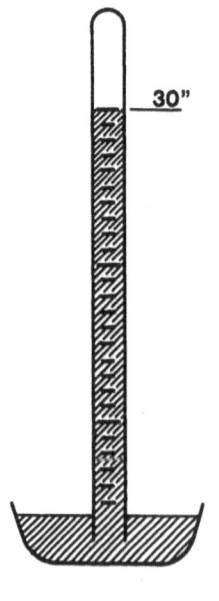

Q: Although air seems like nothing, it can support 200-ton cargo planes—

—and rip buildings apart.

What gives it more substance than we perceive?

A: Air contains an *amazing* number of molecules—300 billion billion per cubic inch. In a space the size of a pinhead, there are millions of times more molecules than there are people on Earth.

The molecules (99 percent nitrogen and oxygen) move ceaselessly at 1,000 miles per hour, and each molecule collides with others five billion times a second.

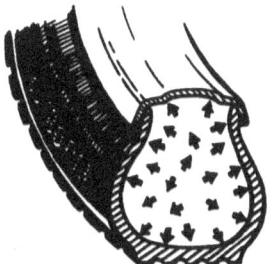

Billions of molecules hitting the inside of a tire provide enough push to support a fifty-ton truck.

Q: Boyle's law says, compress a gas to half the space, and the pressure doubles.

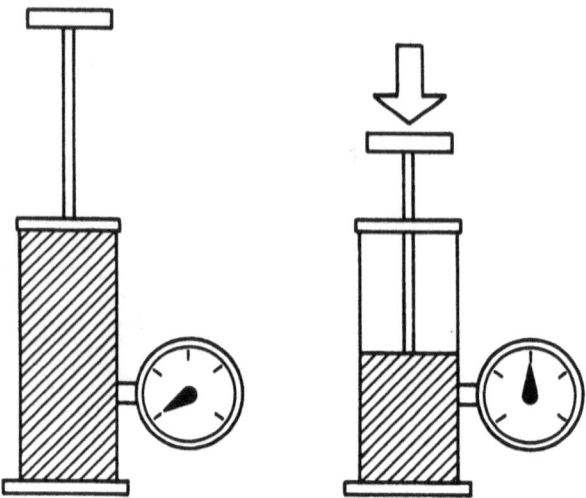

Why does it work with mathematical precision?

A: Because when the same number of molecules are squeezed into a space half as large there are twice as many to hit each portion of wall.

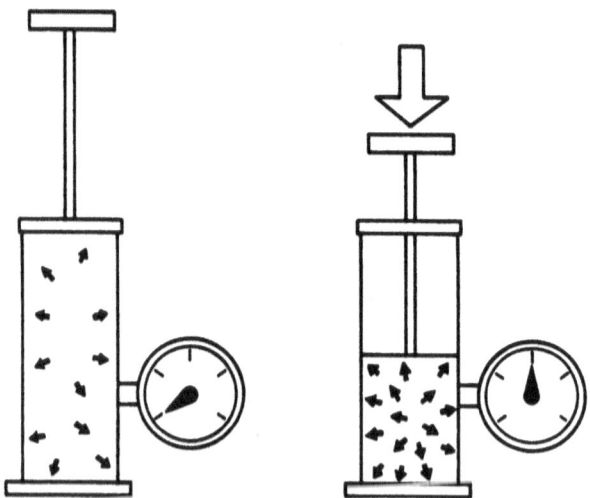

(The law works only at constant temperatures. Sudden compression adds heat, boosting molecular agitation, sending the pressure higher still.)

Q: When you blow above a sheet of paper, it *rises*.

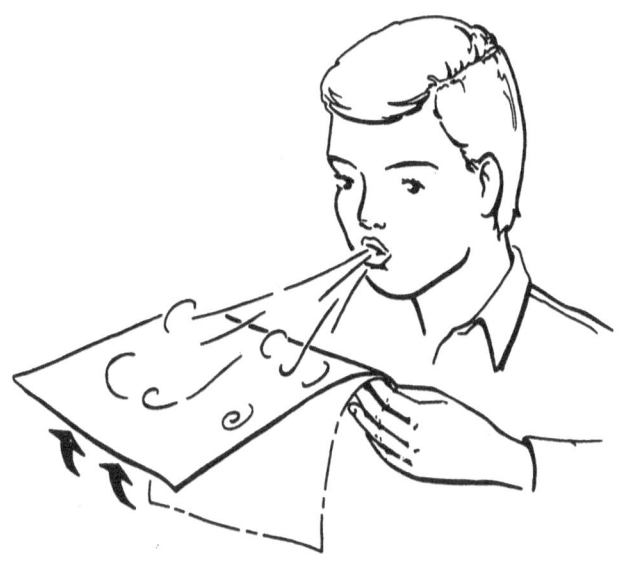

What causes the air pressure below to push it upward?

A: It's the "Bernoulli principle."

Swiss physicist Daniel Bernoulli (1700-1782) discovered that the faster a gas moves, the less pressure it has. Low pressure at a fast-moving spot draws an influx from nearby regions of regular pressure.

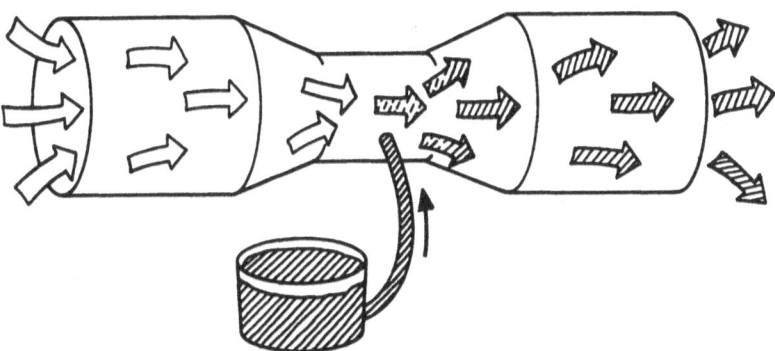

In a carburetor air flows faster in the narrow throat of the venturi tube, causing the gasoline, under normal pressure, to thrust into the low-pressure zone and enter the motor.

Q: What holds an airplane up in the sky?

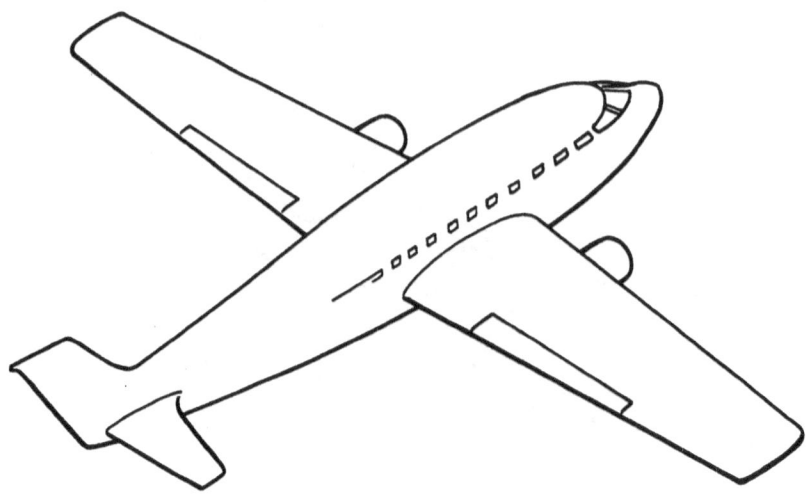

A: The Bernoulli principle explains about 70 percent of the lift.

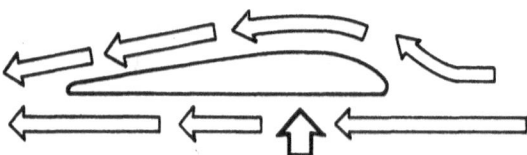

The curved airfoil shape of the top of the wing forces the air to travel farther, hence faster, creating low pressure above the wing. Higher pressure below pushes upward.

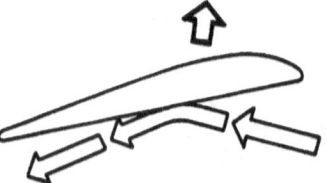

The other 30 percent of the lift is caused by the "angle of attack," which deflects air downward, deflecting the wing up. (Newton's third law of motion: For every action, there is an equal and opposite reaction.) Some high-speed military jets with flat wings rely entirely on this form of lift.

Q: How can a sailboat travel partly *toward* the wind?

A: It's the Bernoulli principle again.

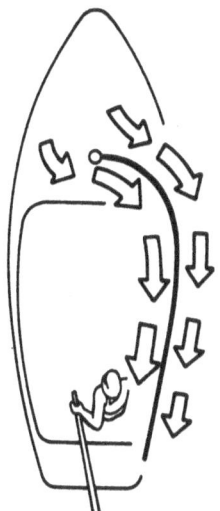

The airfoil curve of the sail forces air to go farther around the convex side, hence faster, creating low pressure in front of the sail. Higher pressure behind the sail shoves forward. (The thrust is partly sidewise, but the bladelike keel on the boat's bottom prevents side-slippage. Caught between the sidewise push and the keel's resistance, the boat is forced ahead, rather like a squeezed watermelon seed.)

Q: The gravitational pull of the moon is the chief cause of ocean tides—but why are there *two* high tides a day, when the moon crosses the sky only once?

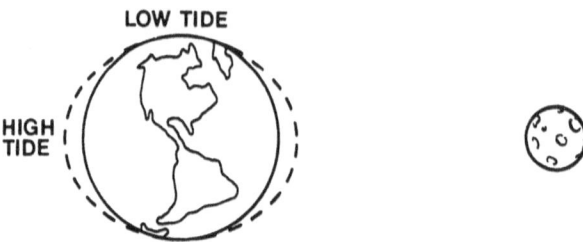

What causes the second oceanic bulge on the side of Earth away from the moon?

A: The moon and Earth actually rotate around a common center of gravity (X) just inside Earth.

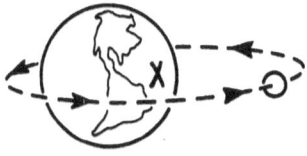

The tidal bulge on the opposite side is caused by inertia from Earth's swing. The water tends to keep moving in the direction of the swing.

The power of gravity fades by the square of the distance, so the moon's pull is much stronger on the near side of Earth and less on the distant side.

Q: Why are tides especially high ("spring tides") during the full moon—

—and during the invisible new moon?

Why do lower "neap tides" occur during half-moons?

A: Because the sun's gravity augments the two pulls at spring tides

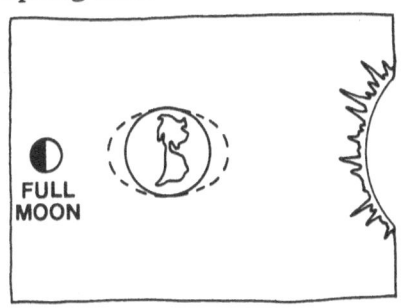

—and counteracts them at neap tides.

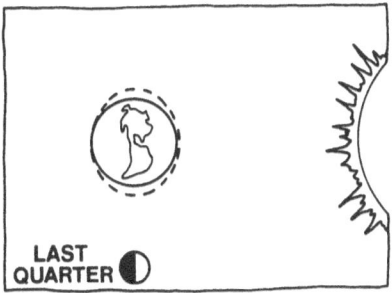

Q: When you see a rainbow—what are you seeing?

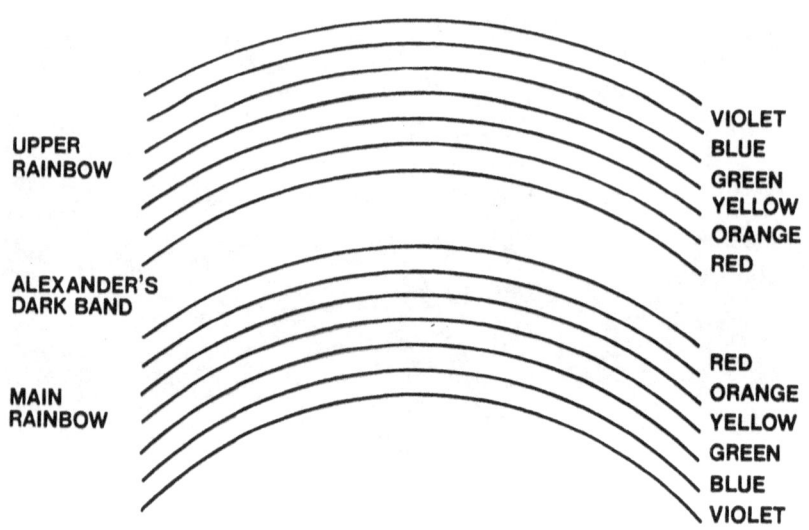

A: You're seeing sunlight split into components with longer and shorter wavelengths. Raindrops act as prisms to separate light.

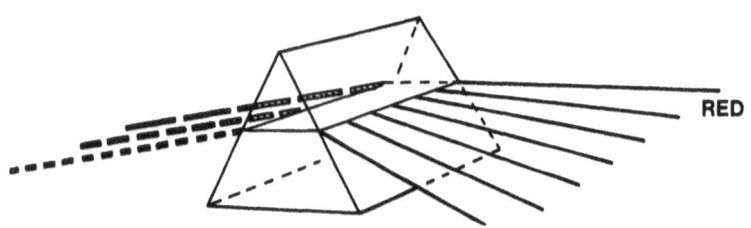

Red has the slowest frequency (midrange: 430 trillion vibrations per second), hence the longest waves (700 billionths of a meter).

Violet has the fastest frequency (midrange: 720 trillion per second) and the shortest waves (415 billionths of a meter).

Q: What *is* light?

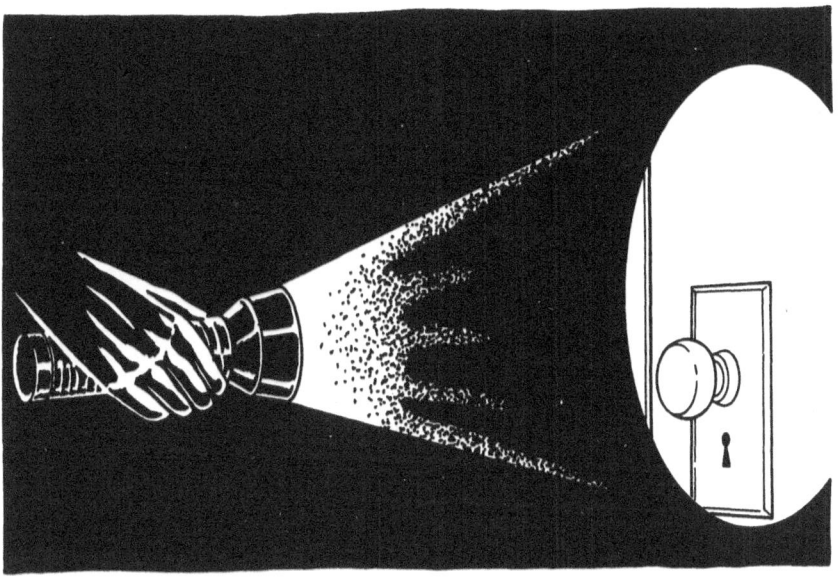

A: One of the universe's profound mysteries.

Albert Einstein showed that light hits in incredibly small massless packets, called photons—

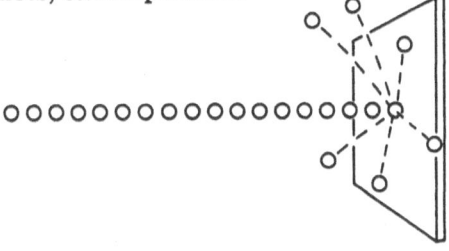

Photoelectric effect: Photons hitting a metal plate dislodge electrons.

—yet light travels in waves.

Thus it's both particle and wave, sometimes jokingly called a "wavicle."

Q: Since visible light is photon waves between 400 trillion and 800 trillion frequency, what about higher and lower frequencies?

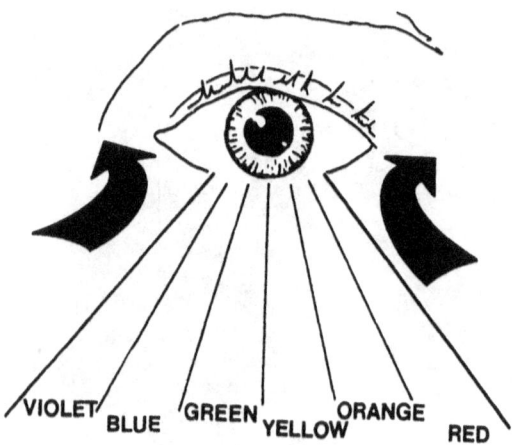

What's in the dark beyond our eyesight?

A: The whole electromagnetic spectrum.

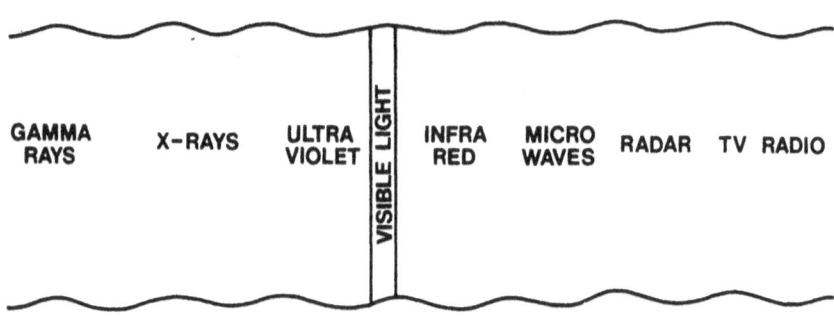

From high-frequency gamma rays with wavelengths of only one-trillionth of a meter, to miles-long radio waves, it's all the same radiation.

Light we see, broadcasts received by our radios, X-rays that reveal our bones—each is photon waves of different frequency, all traveling at the speed of light.

Q: How do polarized glasses control light?

A: The plastic has elongated molecules aligned to form slots, like an incredibly small venetian blind.

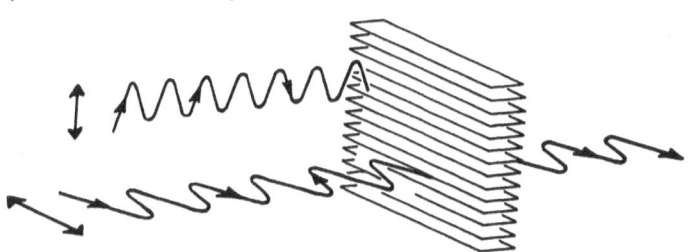

When the slots are horizontal, only horizontal waves get through.

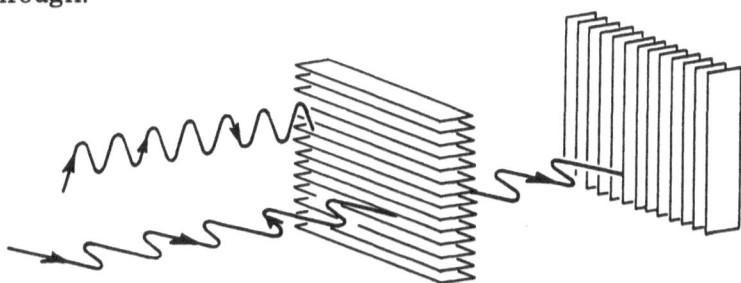

No light penetrates two filters at right angles.

Q: How does parallax measure distances?

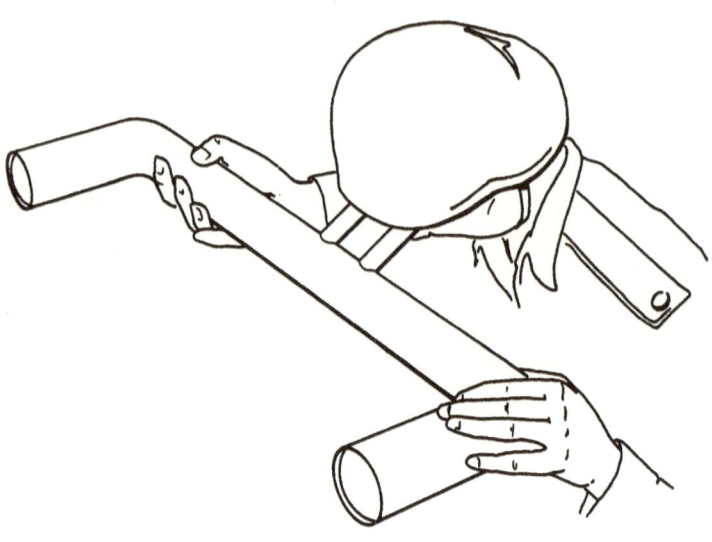

A: From the degree of angle at which two lines of sight converge.

The more the sight-lines vary from parallel, the closer the sighted object.

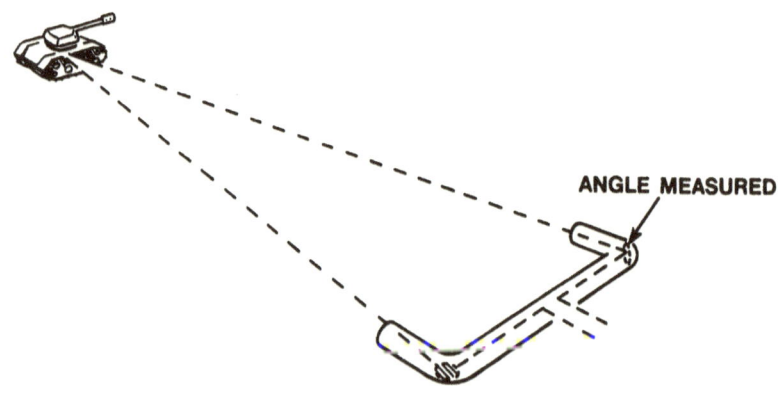

Q: What giant parallax is used by astronomers?

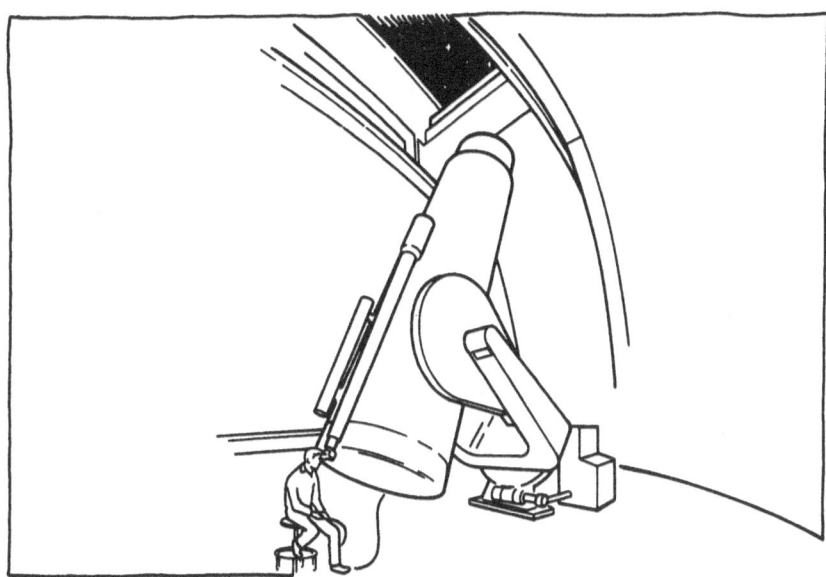

A: Views from opposite sides of Earth—

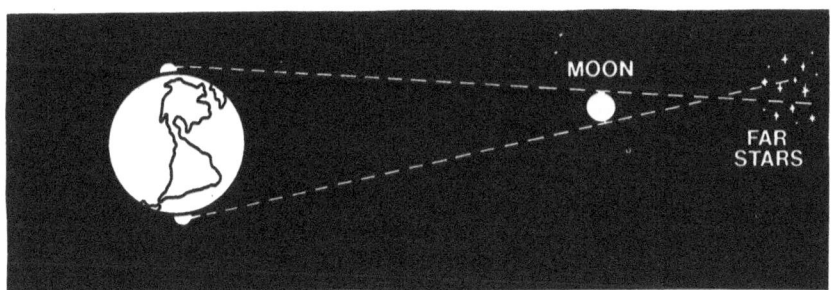

—and from opposite sides of Earth's yearly orbit.

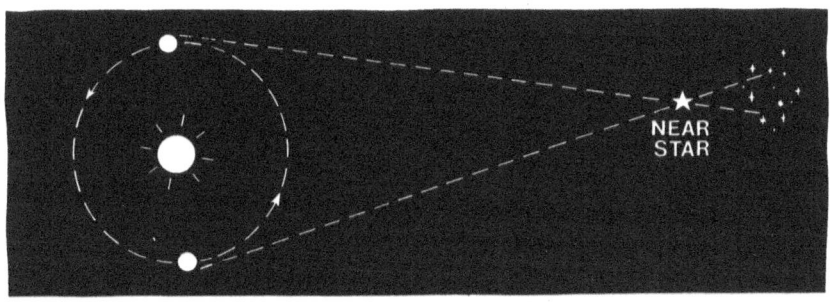

Q: When you see lightning—

—how can you tell its nearness?

A: Count off seconds until the thunder.

Each five seconds equals one mile. Sound travels at 1,100 feet per second—lagging behind light one second for each one-fifth of a mile.

Q: Why can't we hear the terrible roar of the sun's never-ending explosion?

A: Because sound won't travel in the vacuum of space.
Pump the air out of a laboratory jar, and a bell inside becomes inaudible.

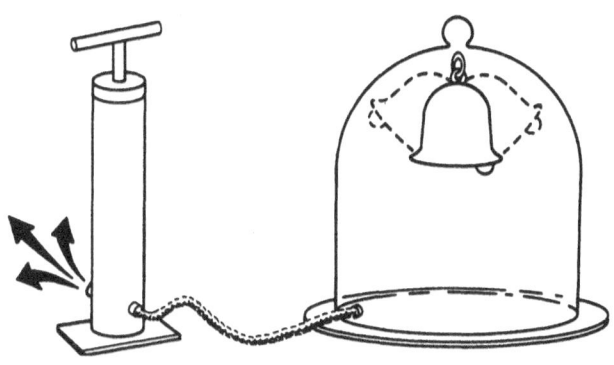

Q: Air around us can be a quivering sea of sound—

—but what *is* sound?

A: It's *compression waves* among the quintillions of molecules in each cubic inch of air. A vibrating sound source pushes and pulls molecules adjoining it.

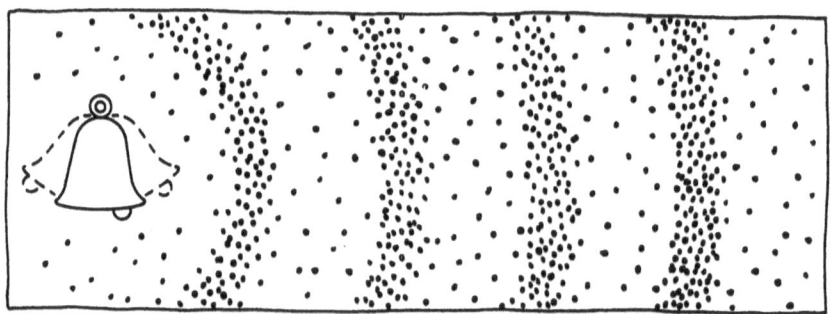

(Example: "Middle C" vibrates 256 times a second. Since sound travels at 1,100 feet per second, C waves are about four feet apart as they fly through an auditorium.)

Q: What are octaves of sound?

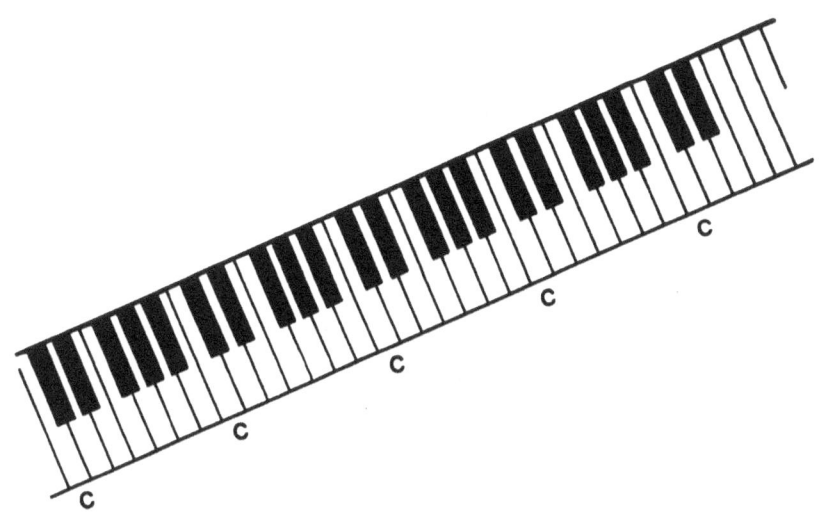

A: The seven-note spans (eight, counting the basic tone at each end of a scale) are a *doubling* and *redoubling* of the frequency. For example, the eight C's of a piano, in one system of tuning, vibrate per second:

C - 4,096
C - 2,048
C - 1,024
C - 512
C - 256
C - 128
C - 64
C - 32

Doubled tones harmonize perfectly, because vibrations of one coincide with alternate vibrations of the other.

Q: Why does a car horn sound *higher* while it's approaching—

—and *lower* after it passes?

A: It's the "Doppler effect," first comprehended in 1842 by physicist Christian Doppler of Prague.

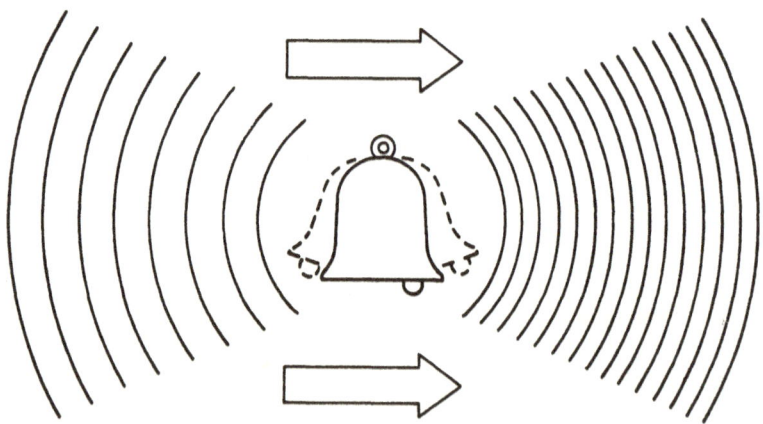

Shorter sound waves make higher tone; longer waves lower tone. When a sound source is moving, it emits each new vibration from a different location—closer to the wave traveling in front, farther away from the wave immediately behind.

Q: How is the Doppler effect used in astronomy?

A: It explains the "red shift," revealing that other galaxies are moving away from us.

Light from those galaxies, when analyzed through a spectroscope (which discloses particular lines for light from different elements), is found to be shifted toward longer red waves. Through the Doppler effect, the galaxies' departing motion "stretches" the waves longer, redder.

Q: What is the chief theory derived from the red shift's evidence that all galaxies are speeding away from each other?

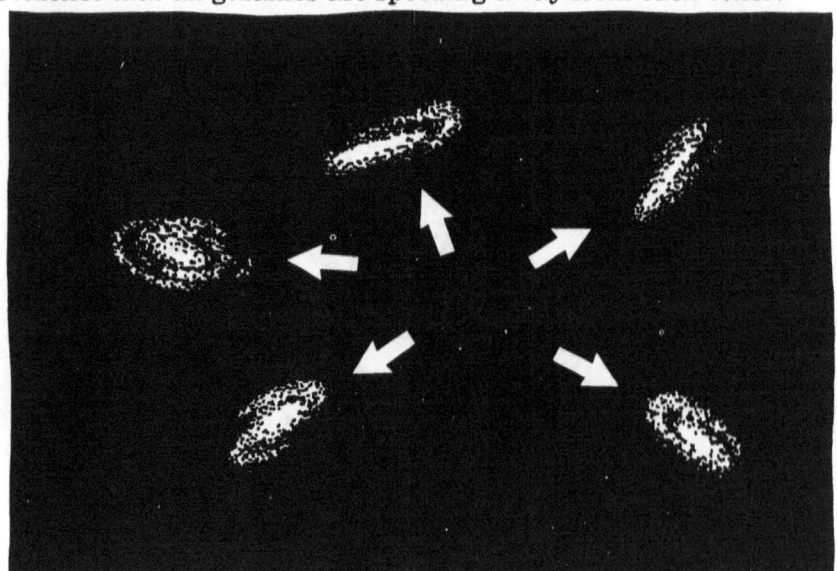

A: The "big bang" theory, which postulates that the universe began with a stupendous explosion 15 to 20 billion years ago.

Now, each second, the visible universe expands by a volume as large as the Milky Way galaxy.

Q: Why does South America's coastline match Africa's?

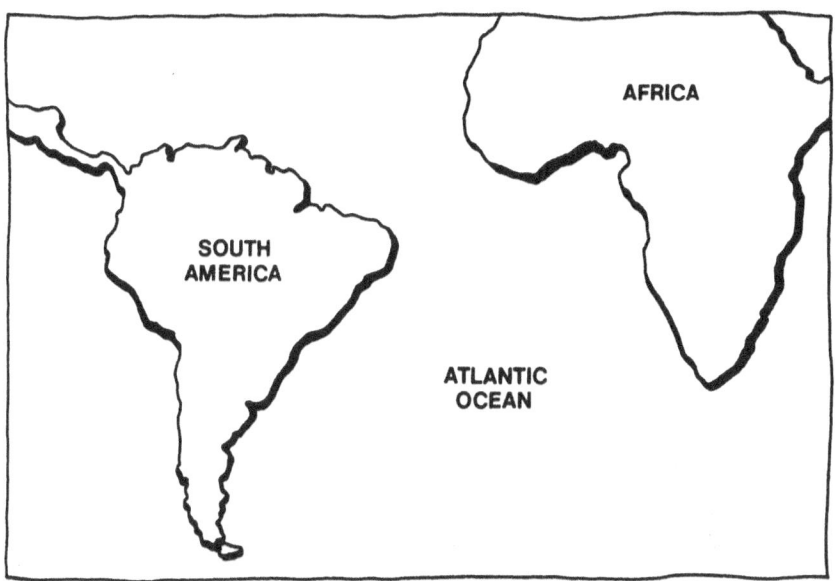

A: Because the Americas once abutted Africa and Europe.

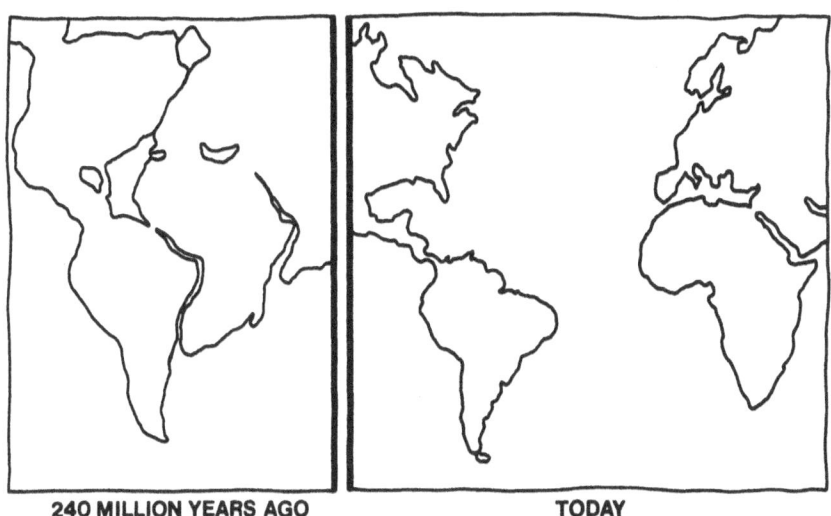

240 MILLION YEARS AGO   TODAY

Q: How can continents—whose weight is beyond comprehension—move?

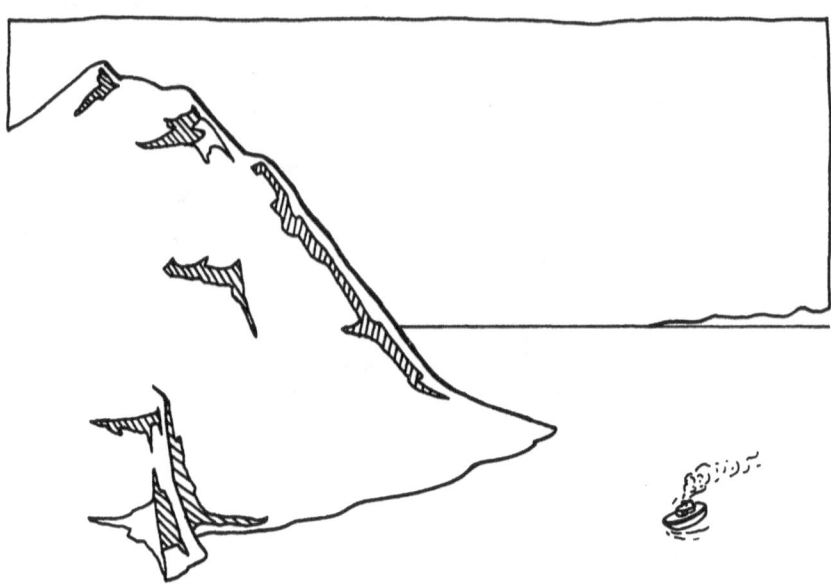

A: Earth's 60-mile-thick crust is cracked into segments (plates) which "float" on the heavier, molten interior.

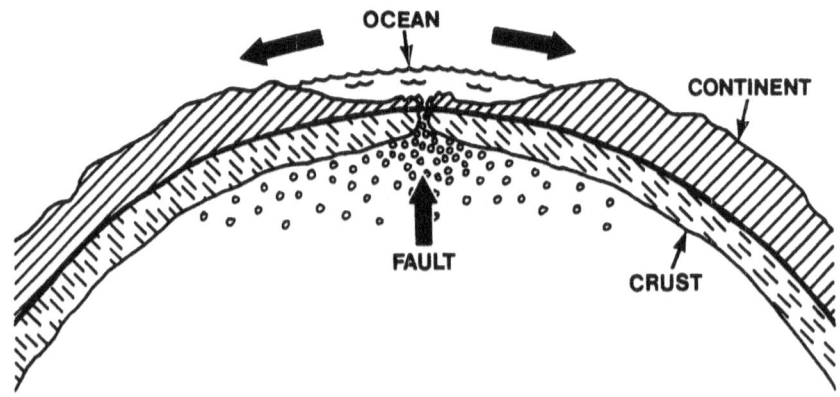

Lava rising through long faults in ocean floors force the plates apart as much as eight inches a year—possibly one mile in 10,000 years or 1,000 miles in ten million years.

Q: What causes earthquakes?

A: Pressure builds between edges of the creeping plates and is released by about one million sudden slippages a year.

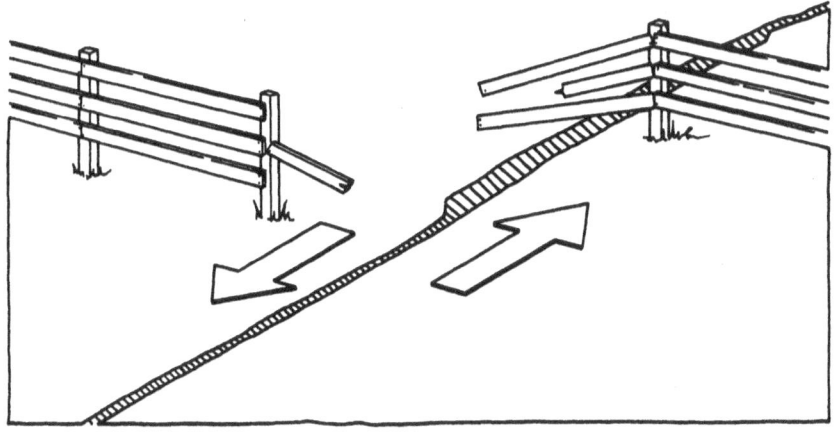

Most of the slips are minor, but some are devastating. During the 1906 San Francisco quake, sections of California jumped twenty feet northward.

Q: What created the Himalayas, the world's highest mountains?

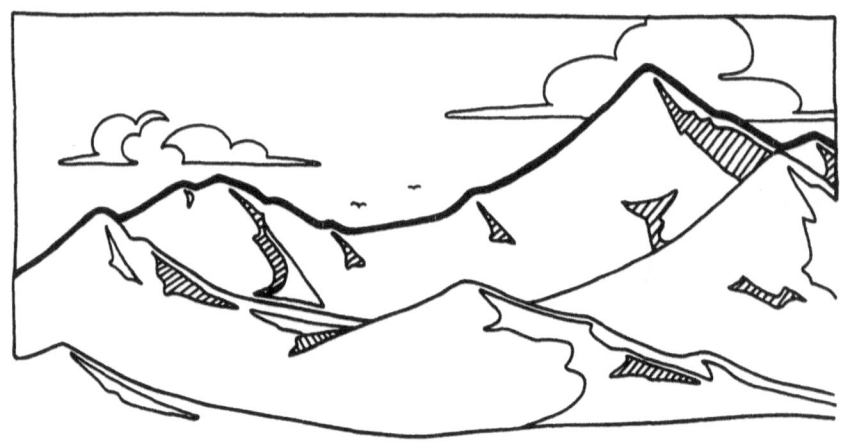

A: The land that is now India once was far south of Asia.

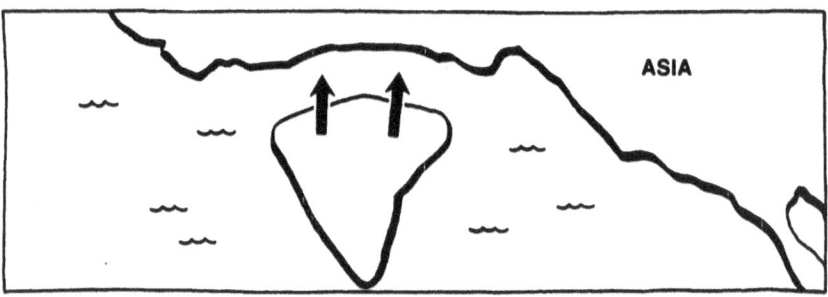

When the Indian plate jammed under the Asian plate fifty million years ago, the force shoved strata nearly six miles high.

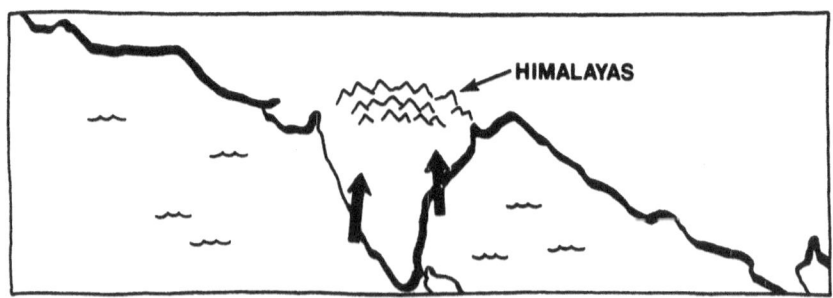

Q: What mountains are even higher, counting underwater portions?

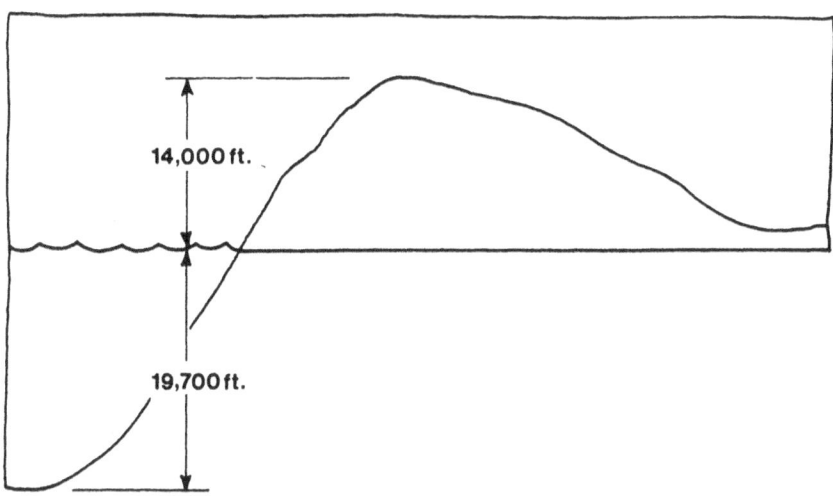

A: The Hawaiian Islands were created over a volcanic "hot spot" that burned through the ocean floor periodically.

The lava climbed nearly 20,000 feet to sea level, then 14,000 feet higher.

Gradual northwest movement of the Pacific plate arranged the islands in a string.

Q: When you're in a limestone cavern—

—where are you?

A: At the bottom of a prehistoric sea.

Remains of millions of shellfish were compressed under layers of silt and hardened into limestone.

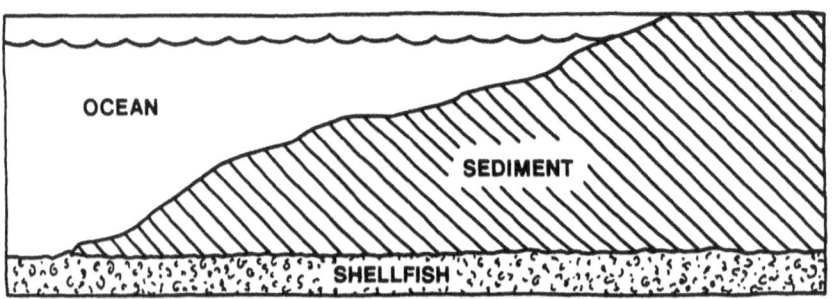

Eventually the rock was forced upward, sometimes high in mountains. Then underground water slowly dissolved the lime and made a cavern.

Other seashell material, compressed more intensely, became marble.

Q: Do you realize that prehistoric swamps and ocean beds operate your television set—

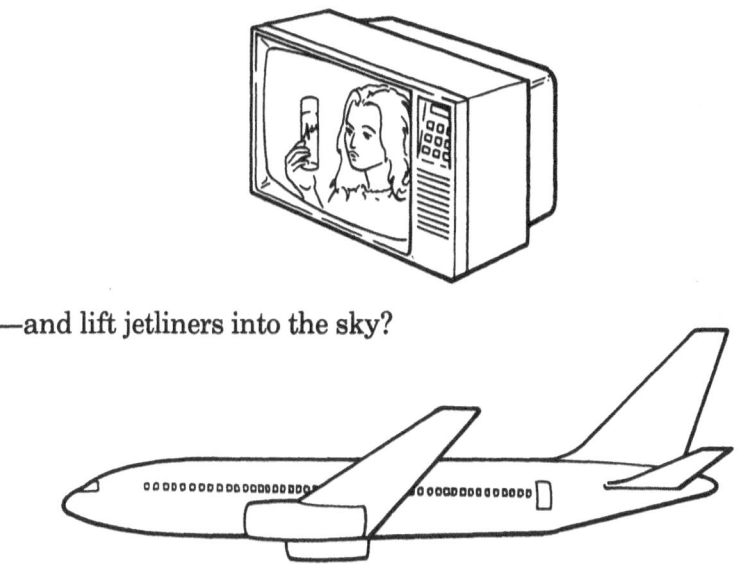

—and lift jetliners into the sky?

A: Dense jungles of the Carboniferous period (345–280 million years ago) sank into beds that were compressed into coal.

And organic beds of marine plants and animals followed a similar process to become oil and gas.

Now they generate electricity, propel motors, heat homes, run factories, and meet other energy needs.

Q: Isn't all of Earth's energy "solar energy"?

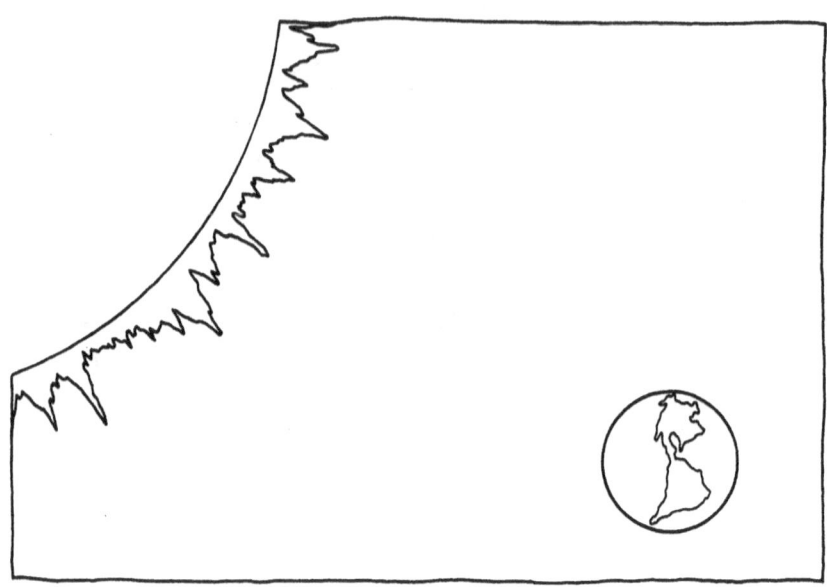

A: Mostly, yes. Sun rays heat the land unevenly, causing most winds. They evaporate water, making the water cycle that powers hydroelectric dams. Sun power grew all the vegetation that became coal, oil, gas, and peat.

Sitting before a fireplace, a wise scientist told his daughter that the crackling fire was "stored-up sunlight unwinding from the wood."

However, the energy in volcanos and atoms is innate to Earth's own material.

Q: Speaking of fireplaces, do you realize that logs don't actually burn?

What is fire?

A: Only gas burns. Flame—the self-sustaining chemical reaction of rapid oxidization—doesn't occur in solids or liquids.

When a combustible substance becomes hot enough (the ignition point), it emits hot gas that reacts with oxygen gas in adjoining air. Heat from the flame drives out more gas, keeping the reaction going. The remaining ash is minerals that wouldn't vaporize.

(Slower oxidization in your cells powers your body and keeps you warm. Still-slower oxidization turns iron into rust.)

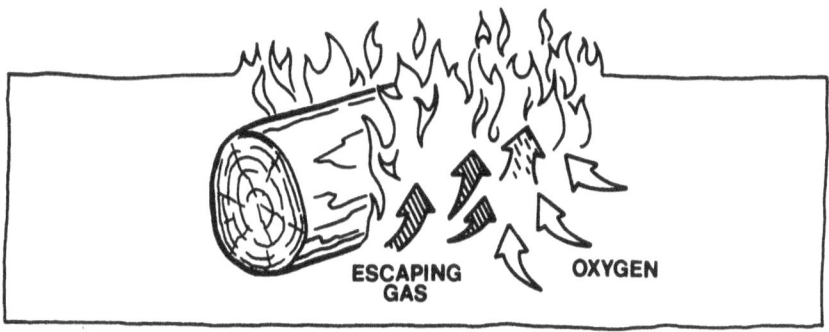

Q: Why do clouds float at a particular level in the sky?

A: They're continuously created at that level by atmospheric coldness.

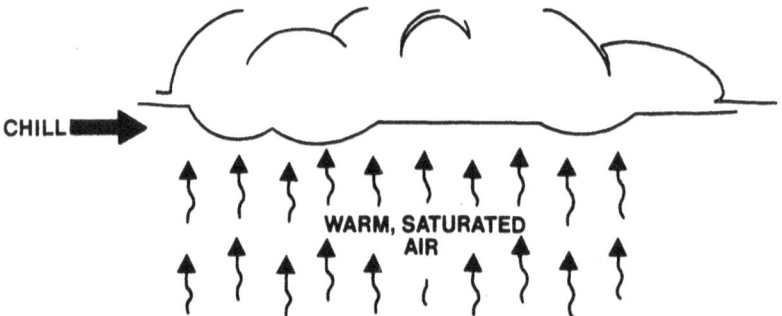

Warm air can hold more invisible water vapor than cold air. It becomes saturated to its maximum capacity. Since warm air expands, it is lighter and rises. When it reaches a chilling altitude, it is unable to hold the vapor, which condenses into visible microscopic droplets.

Q: Most rain that falls from clouds runs off into rivers—right?

A: Wrong. Only about one-fifth runs off. Another fifth goes into the underground "aquifer" strata.

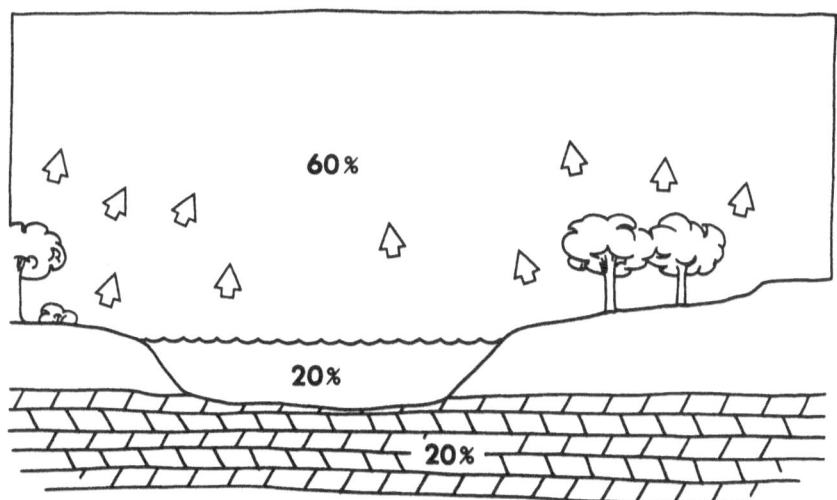

By far the largest portion is evaporated by sunshine or *transpired* through plant leaves. An acre of corn transpires up to 4,000 gallons of water a day.

Q: What is the world's biggest transport system?

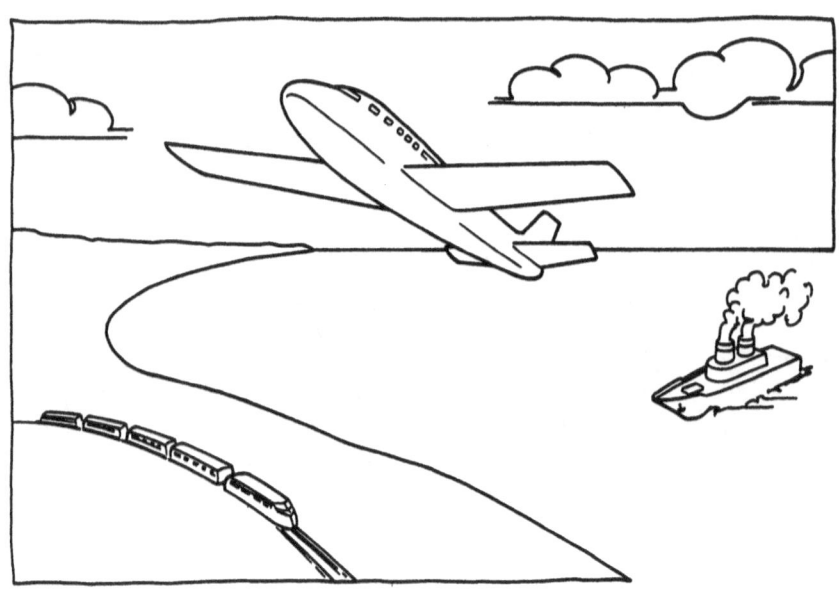

A: The air. It weighs 5.6 quadrillion tons and carries immense loads.

A single cloud can carry up to 1,000 tons of moisture.

Air draws 5.5 billion gallons of water an hour from the Gulf of Mexico on a hot day and "airlifts" it over the U.S. Northeast by millions of tons.

Q: Where is wind four times stronger than a hurricane?

A: From five to eight miles above Earth, jet streams blow as severely as 400 miles an hour (giving a fuel-saving lift to high-flying planes).

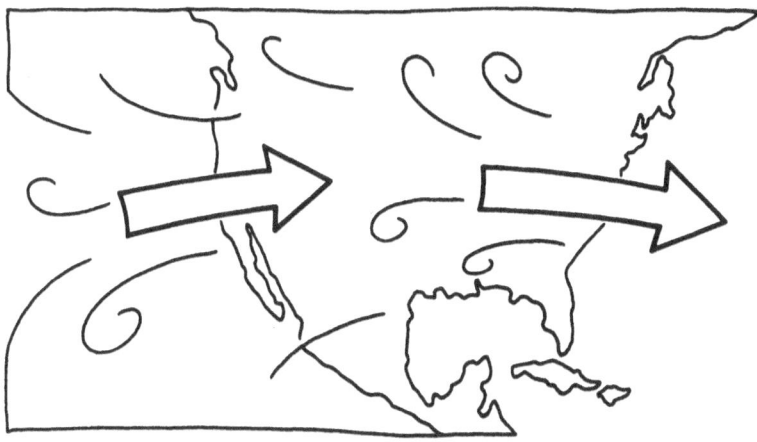

The streams flow from west to east, above the fronts where cold and warm air masses meet.

Q: Air grows steadily colder at higher altitudes—right?

A: Wrong. It gets colder for many miles, reaching a bitter -135°F at fifty miles up.

But above 100 miles, where merciless solar radiation splits gas molecules into charged ions, temperatures are 2,000 to 3,000°F. This is the "thermosphere."

However, particles are extremely sparse at that altitude, so the fiery ions have little power to heat passing objects.

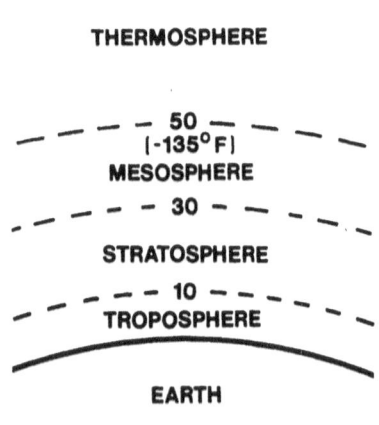

Q: Perpetual motion is an unattainable dream for inventors—yet it's all around us, everywhere.

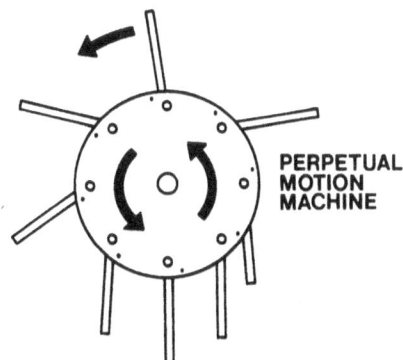

Even in utter stillness, there is ceaseless motion.

How so?

A: Besides the visible perpetual motion—orbiting planets, flowing rivers, winds, tides—there is invisible perpetual motion in atoms and molecules.

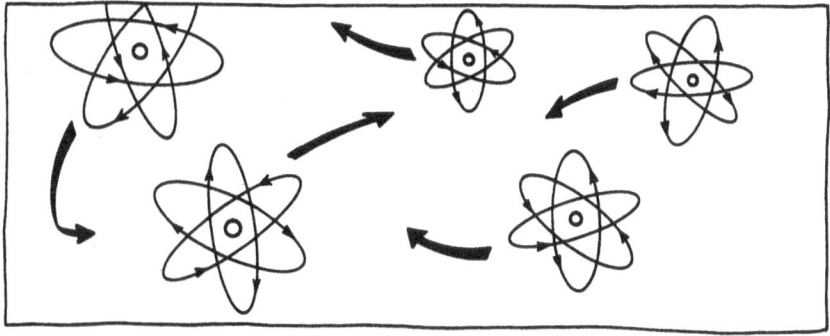

The nucleus seethes in ferment. Electrons whirl so frantically they make a solid-seeming "cloud." Whole atoms, locked in the crystals of solids, buzz with vibration. Molecules (combinations of atoms) quiver in solids, slide in liquids, and fly frenetically in gases. In metals, free electrons dart at two million miles per hour.

Q: Everyone knows the universe is composed of atoms—

—but can the human mind grasp the incredible smallness of atoms?

A: Only with great difficulty.

If a baseball were enlarged to the size of Earth, its atoms would be the size of grapes.

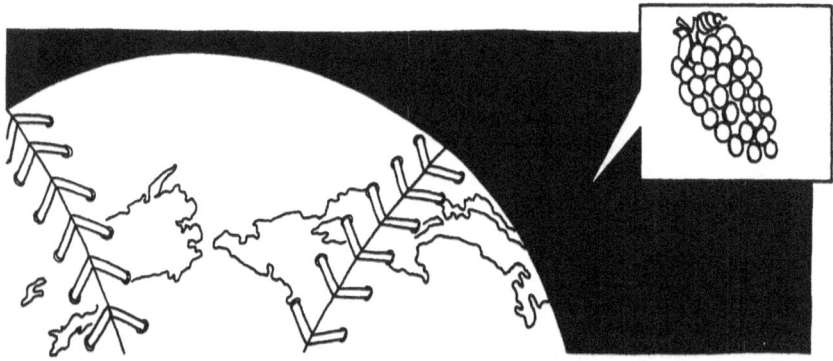

If atoms were as large as pinheads, the atoms in a single grain of sand would fill a one-mile cube.

One drop of water contains 100 billion billion hydrogen and oxygen atoms in 33 billion billion H_2O molecules.

Q: How much smaller are the particles inside atoms?

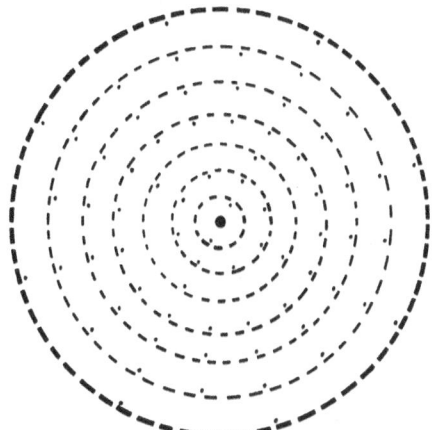

A: Astoundingly, they're as much smaller than atoms, as atoms are smaller than us.

If an atom were the size of a fourteen-story building, its nucleus—the heart, containing virtually all its mass—would be the size of a *grain of salt*.

Like the solar system, an atom is mostly vast emptiness.

Q: Under the newest, strongest microscopes, some atoms look like spheres.

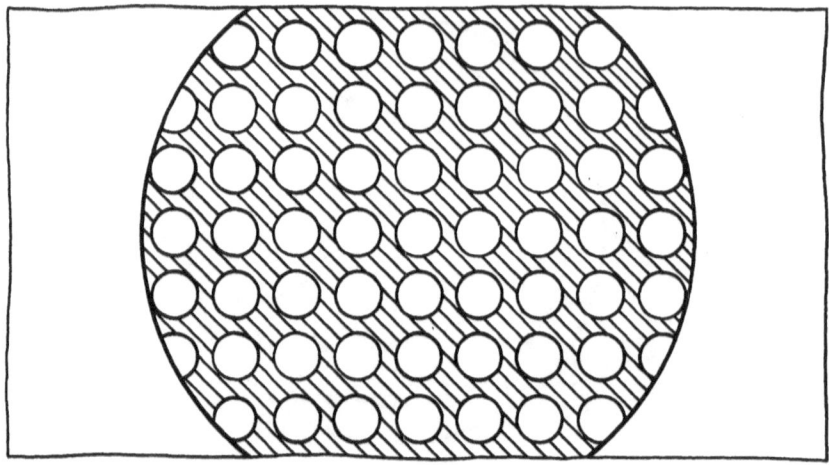

Gold atoms locked in their crystal lattice—seen at 25 million magnification.

How can they look like spheres if they're 99.999 percent void, as empty as the solar system?

A: Because electrons buzz around the nucleus at *fantastic* speeds—billions of whirls in a millionth of a second.

Their blur of movement—seemingly being everywhere at once—gives form to the atom. (It's somewhat like the way a whirring airplane propeller seems to be a disk, not a blade.)

Q: Since similar charges repel each other, how can an atom's positive particles, the heavy protons, coexist in the nucleus?

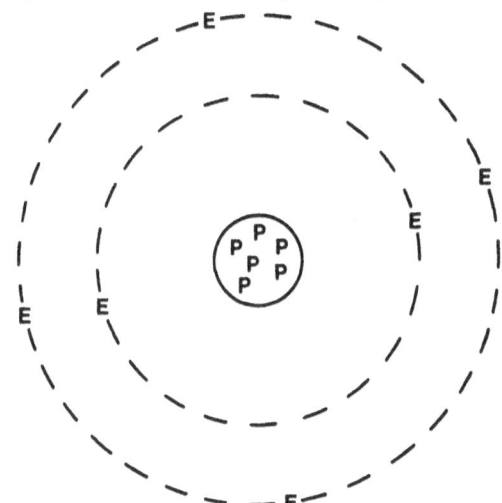

Carbon atom: six positive protons balancing six negative electrons.

A: Other heavy particles, neutral neutrons, possess the ability to bind protons into a stable nucleus.

Carbon 13: seven neutrons hold six protons.

This nuclear binding—the strongest force in nature—is accomplished by the rapid exchange of tiny pions, or "gluons," between protons and neutrons.

Q: What are isotopes?

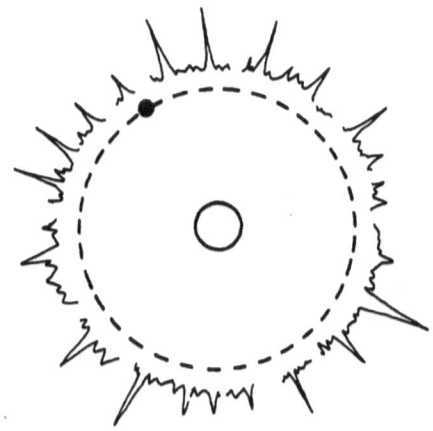

A: Different forms of an element, produced by varied numbers of neutrons in the nucleus.

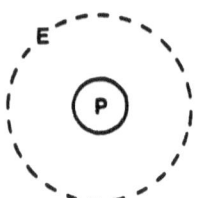

Hydrogen, the lightest element, usually has one proton to match one electron—

—but "heavy hydrogen" isotopes called deuterium and tritium have neutrons. Their weight is heavier, but their chemical properties remain nearly unchanged.

Q: Why are an atom's electrons usually visualized in concentric spheres?

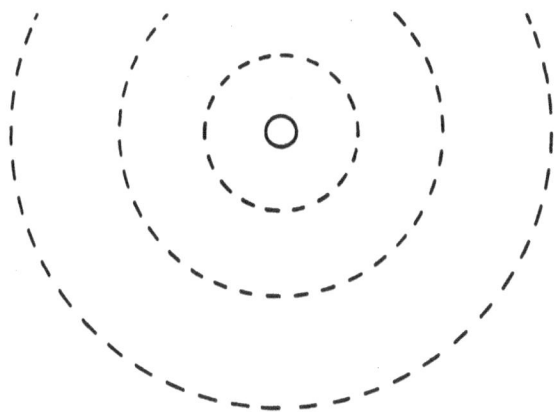

A: Electrons inhabit "shells." The innermost shell has a capacity for 2 electrons, the second 8, the third 8. Heavy atoms have shells of up to 32 electrons—but the outer one nearly always has a limit of 8.

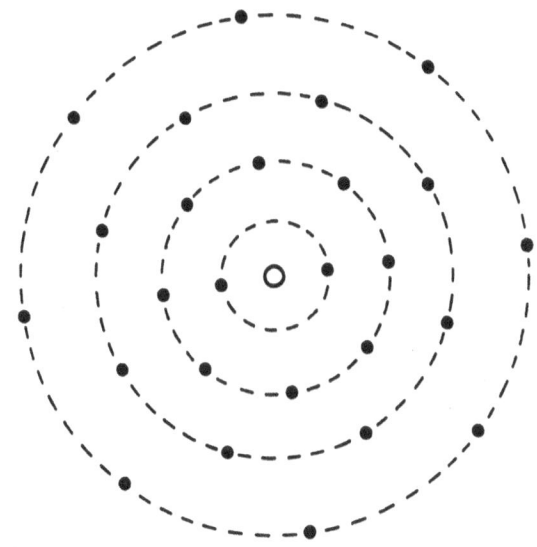

Q: What causes atoms to join into molecules, forming chemical compounds?

A: The outer shell of electrons governs chemical activity. Molecules become stable when these shells are filled to capacity.

Atoms unite and link together *electron pairs* (always two electrons with opposite spin), so the shared electrons will satisfy the number needed (the valence) to finish a shell of 2 or 8 in each atom.

Atoms with just enough electrons to fill a shell are inert and usually won't form molecules.

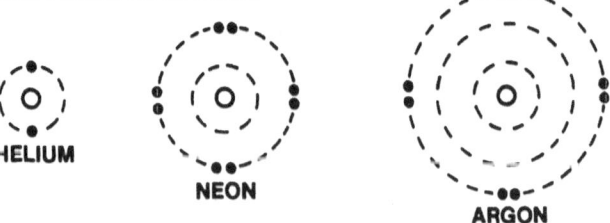

Q: What makes some molecules so durable that they survive unchanged for millions or billions of years?

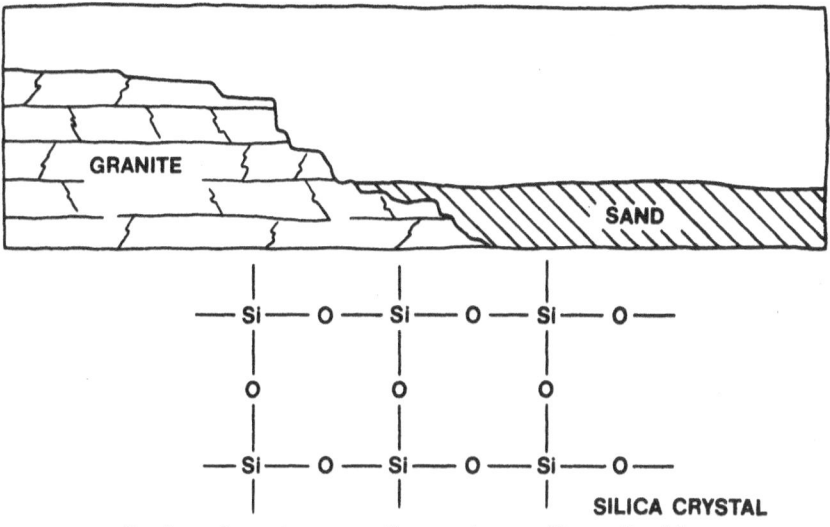

Rock and sand are mostly quartz, or silicon dioxide, SiO_2—silica.

A: The electrical bond between atoms is strong and permanent, unless altered by a new chemical reaction.

There are two main types of bonds:

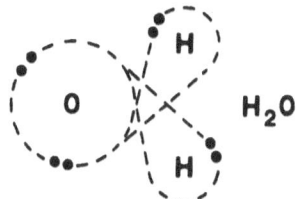

"Covalent bonds" occur when combined atoms share electrons.

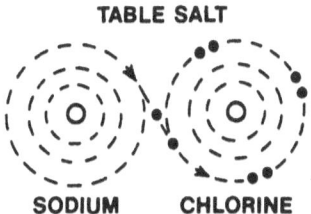

"Ionic bonds" occur when an atom needing an electron or two seizes them from an atom with one or two in its outer shell. Thus each becomes a charged *ion*—one negative, one positive—and their opposite electrical attraction holds them together.

Q: Since atoms combine to complete outer shells of 2 or 8, do chemical properties of elements follow a pattern of 8's as the elements increase in atomic number (the number of protons, or electrons, per atom)?

<pre>
        Hydrogen—1   Sodium—11
        Helium—2   Magnesium—12
        Lithium—3   Aluminum—13
        Beryllium—4   Silicon—14
        Boron—5   Phosphorus—15
        Carbon—6   Sulfur—16
        Nitrogen—7   Chlorine—17
        Oxygen—8   Argon—18
        Fluorine—9   Potassium—19
        Neon—10   Calcium—20
</pre>

A: Yes. In 1869, before the electrical structure of the atom was discovered, Russian chemist Dmitri Mendeleev perceived the recurring pattern and devised the Periodic Table of Elements.

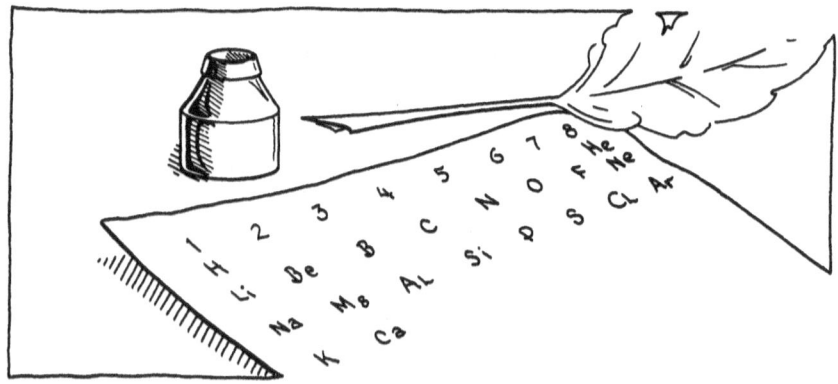

Beyond calcium, heavier elements have extra electrons in some shells. Their molecule-combining involves two or more outer shells, and the periodic pattern is complicated.

Q: If atoms and molecules are as empty as the night sky containing the planets of the solar system—

—how can steel, rock, wood, plastic, and other substances composed of these voids be solid?

A: Solidity is an electrical illusion.

The electron clouds, charged negative, repel each other with a firm "push."

In most solids, atoms and molecules are locked into crystal lattices. They're held together by various electrical bonds—and simultaneously they're held faintly apart by repulsion of the negative electrons.

They hover in place, not touching, but quivering as though suspended on invisible springs.

Q: How can so few subatomic particles and so few types of atoms—

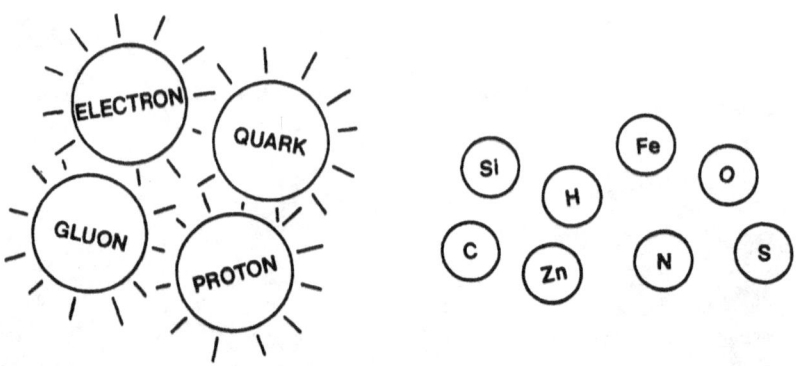

—produce the incredible diversity of our world?

A: Remarkable transformations occur as the basic units combine.

For example, when carbon atoms bind onto themselves in six-sided rings, they become sooty graphite (pencil lead)—but when they bind into tetrahedrons, they become dazzling diamonds.

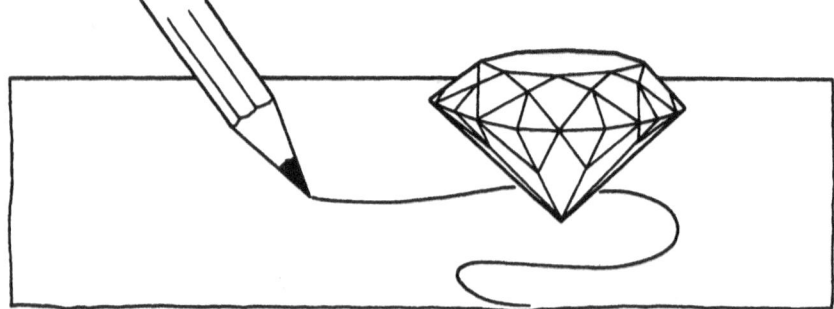

Q: Why does a substance exist in three states—solid, liquid, and gas?

A: The key is heat.

In a solid, the vibrating molecules are too cold to overcome the cohesive forces binding them into crystals.

Increased heat agitates the vibration until they break the crystal and slide around each other. They have passed the *melting point* and become a liquid.

Still more heat imparts greater motion until the molecules break the liquid's surface tension and escape. They have passed the *boiling point* and become a gas.

Q: Is there no halt to the eternal whirling, buzzing, spinning, colliding, and darting in the realm of atoms and molecules?

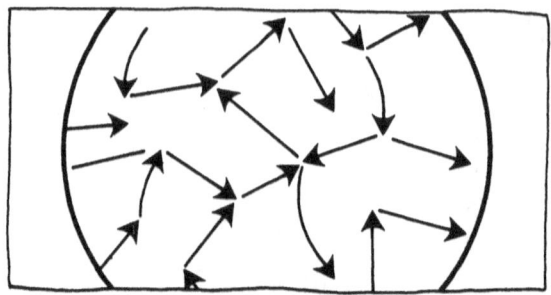

Brownian motion—arrows represent smoke particles hit by molecules.

A: At Absolute Zero (-273°C or -460°F)—the point totally devoid of heat—all matter is frozen solid, and motion *nearly* ceases. Yet some vibration remains in the atoms locked in crystals.

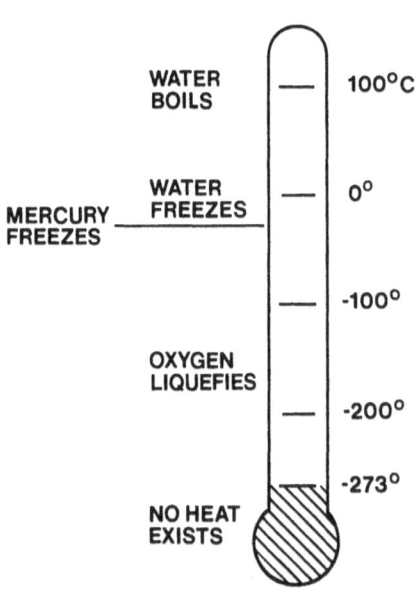

Q: Although atoms are indestructible and eternal in most circumstances—

—what are two awesome exceptions?

A: Two nucleus-altering reactions: *fission* and *fusion*.

Each follows Albert Einstein's historic equation E=MC², which indicates that stupendous power is released when matter is converted into energy.

(Energy equals the mass of the matter multiplied by the constant, the speed of light, squared.)

Q: How does fission work?

A: When a neutron hits the nucleus of a heavy, unstable element like uranium 235, the nucleus splits into lighter elements, releasing part of the "binding energy" plus two or three more neutrons which hit other nuclei in a rapid *chain reaction.*

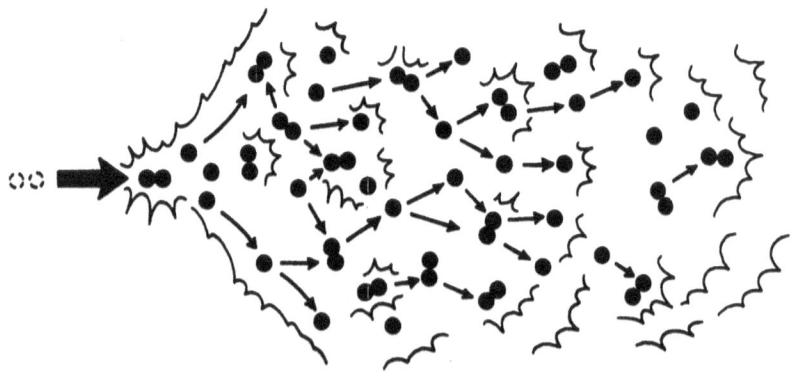

Only one-thousandth of the uranium is converted into energy.

In the bomb that killed 140,000 people at Hiroshima in 1945, a quantity of matter smaller than a *dime* turned into energy.

Q: How does fusion work?

A: Fusion—the process that occurs in stars and hydrogen bombs—merges hydrogen nuclei to form helium and heavier elements. Part of the mass is lost during this reaction. For bombs, heavy hydrogen isotopes are used.

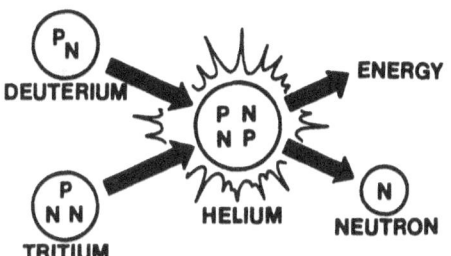

Fusion converts a much larger portion of the mass into energy than fission does.

The reaction won't occur without colossal heat and pressure such as exist inside a star. To trigger a fusion bomb, a fission bomb is detonated to create starlike conditions.

Q: In other words, the 200 billion stars of the Milky Way galaxy—and those of ten billion other observable galaxies—are continuous hydrogen bombs.

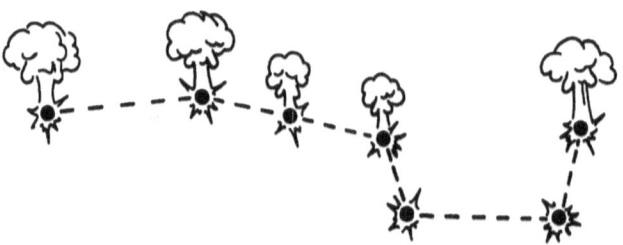

What makes stellar fusion occur?

A: A star's mass is so great that its enormous gravity compresses atoms until their electron orbits are crushed, producing a dense "plasma," sometimes called the fourth state of matter.

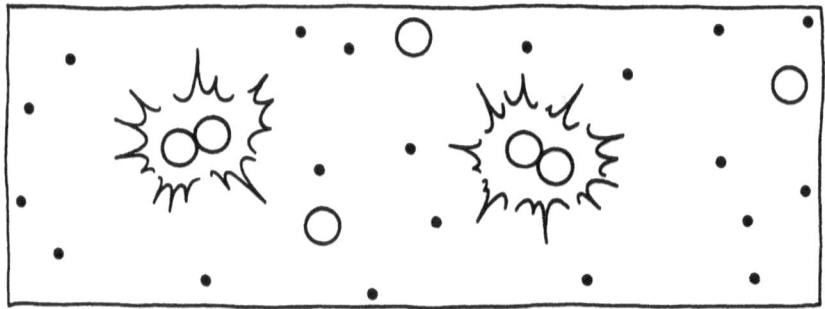

Without shells of electrons buffering them, nuclei crash together and fuse into heavier elements.

Q: In what other ways does the smallest realm, the inside of the atom—

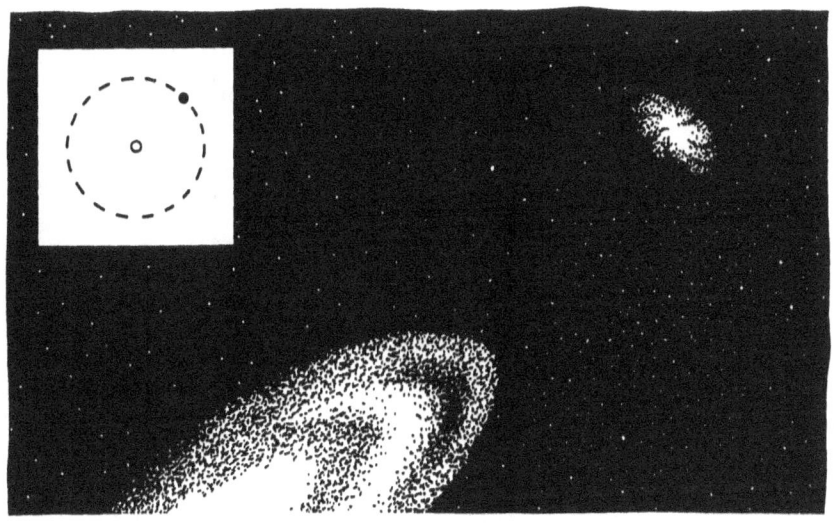

—govern the largest realm, the universe?

A: Subatomic behavior, under horrendous gravity, produces three mind-boggling space wonders: white dwarfs, pulsars (neutron stars), and black holes.

Their incredible properties derive from loss of empty space between subatomic particles.

Q: What are white dwarfs?

A: Tiny stars 10,000 times denser than steel. Their material weighs ten tons per thimbleful. (Can you imagine a thimble so heavy that 100 strong men couldn't lift it or a tractor couldn't pull it?) They occur when a star about the size of our sun burns low and collapses upon itself.

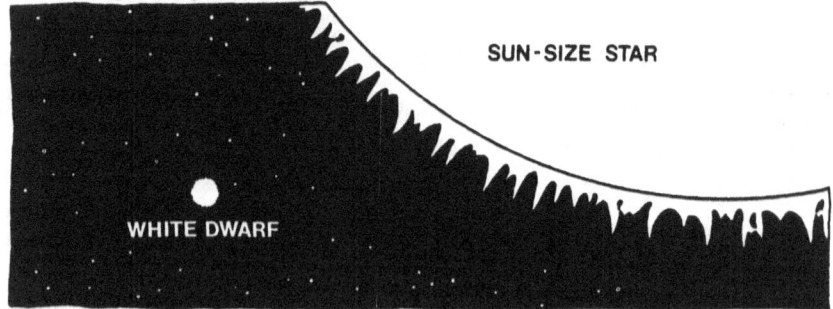

The plasma "soup" is compressed to its maximum, a point at which resistance by electrons halts the squeeze of gravity.

Q: What are pulsars (neutron stars)?

A: The second stage of collapse.

A teen-age genius, Subrahmanyan Chandrasekhar, computed that, if a collapsing star has 1.4 times the mass of our sun, its gravity is too great for the electrons to halt the compression.

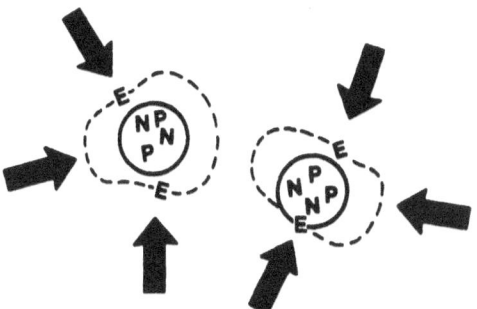

They're crushed into the nucleus, where they merge with protons to create a solid mass of neutrons. This material weighs *10 million tons per cubic centimeter*—more than the mind can comprehend.

Such neutron stars rotate several times a second, emitting a radio pulse with each rotation.

Q: What are black holes?

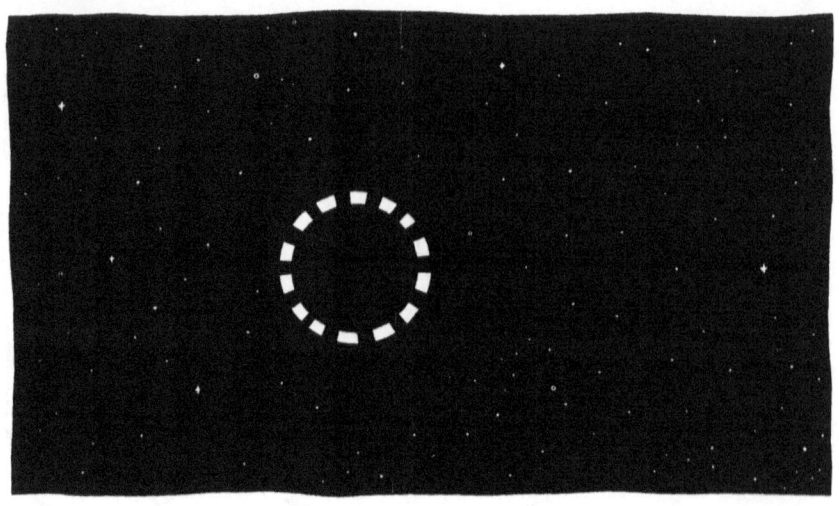

A: The final, incredible state of collapse.

If the star's remains are three times larger than our sun, the gravity is too great to be checked at the neutron star level.

The overwhelming gravity pulls all matter, even whole stars, into a vanishing "pit of infinity." Not even light can escape.

Planet Earth would be no larger than a *pearl* if it were compressed to the threshold (the Schwarzschild radius) at which a black hole begins. Actual size is shown in the inset.

Q: If neutron star material, the "solid" parts of atoms packed together without their empty space, weighs ten million tons per cc—

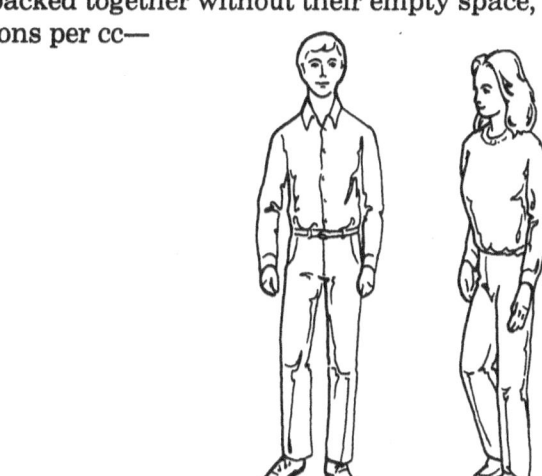

—how much actual matter is in a 180-pound man or a 120-pound woman?

A: Too little to see with a microscope—which defies belief.

Seeming impossibilities like this caused Scottish scientist J.B.S. Haldane to say, "The universe is not only queerer than we suppose, but queerer than we *can* suppose."

Q: What are quasars?

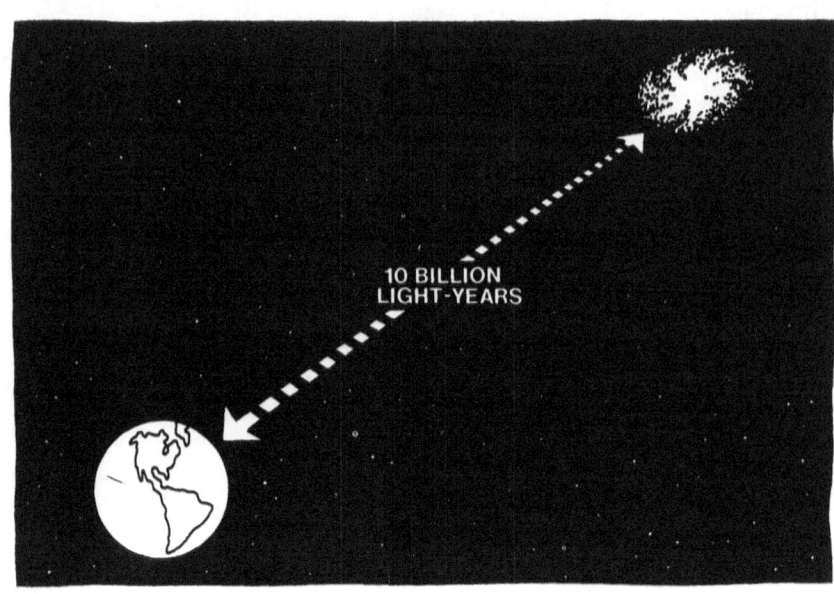

A: The most distant known objects in the universe—and the most powerful. They emit the energy of hundreds of galaxies. They are receding from us at 90 percent of the speed of light. The light we see left them about ten billion years ago, when the universe was younger.

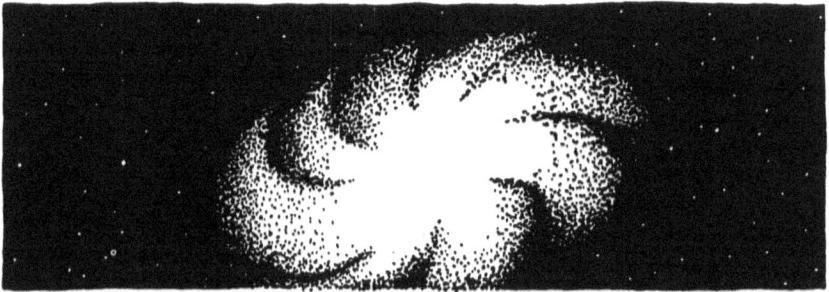

Theory: When a galaxy is young, a monster black hole at its center swallows nearby stars, and their death throes produce terrible torrents of radiation. Later, when the black hole consumes all "food" within reach, it becomes quieter and the quasar phase ends.

Q: Why do scientists say that people, trees, houses, and hills are made of "star stuff"?

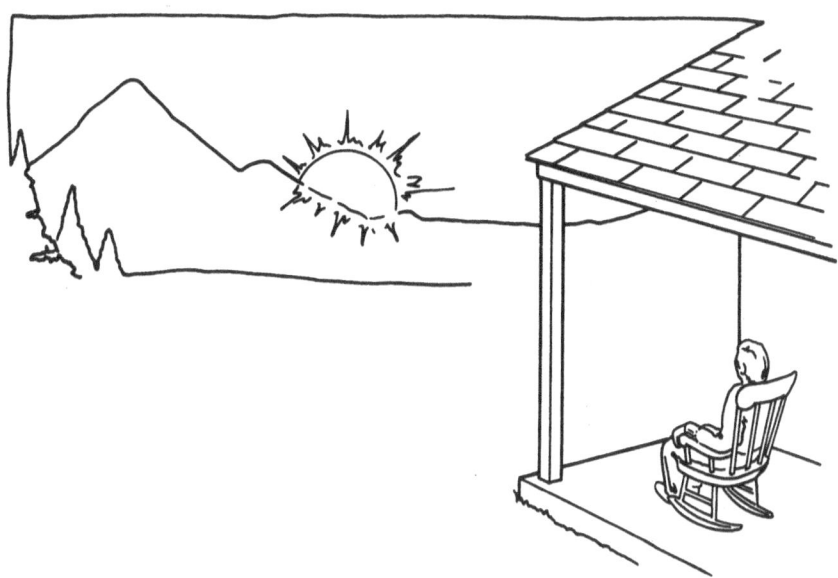

A: Computations indicate that the original "big bang," which flung the universe into existence, created only hydrogen and helium. Stars and galaxies condensed from these light gases.

Most heavier elements now composing our world were created by fusion inside stars—and the heaviest of all were fused when stars exploded as supernovas.

Q: Since atoms are electrical, everything in the universe is electrical.

How do atom charges—negative electrons and positive protons—become the various types of electricity we know?

A: Travel by electrons—that's electricity.

When these negative particles become detached from their atoms, they carry their charge along, and remnants of the atoms become positive-charged ions.

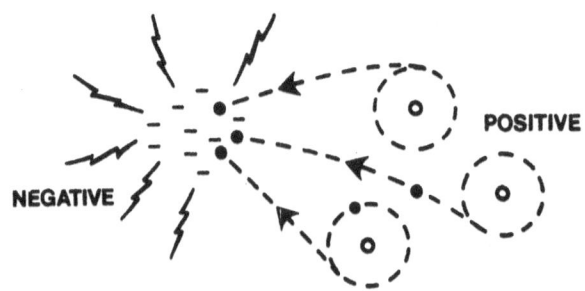

Static (stationary) electricity occurs when electrons leave one substance and cling to another.

(An exception: In liquids and gases, electric current can consist of traveling ions.)

Q: How does electricity flow through a wire at the speed of light?

A: *Conductors* are substances already teeming with "free electrons" darting randomly as fast as two million miles per hour.

When a current enters, these electrons surge.

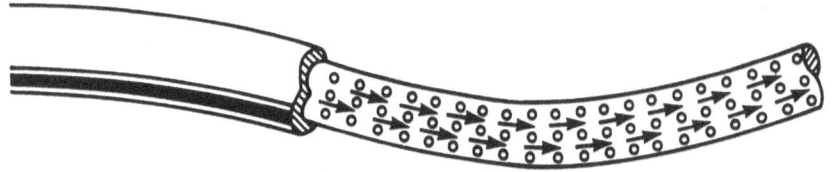

They don't move along the wire at the speed of light, but the initial impulse does—rather like the way a full garden hose begins squirting from its nozzle the moment pressure enters the other end.

In a sixty-watt bulb, three billion billion electrons per second pass through the filament.

Q: Why won't electricity flow through rubber, plastic, and other insulators?

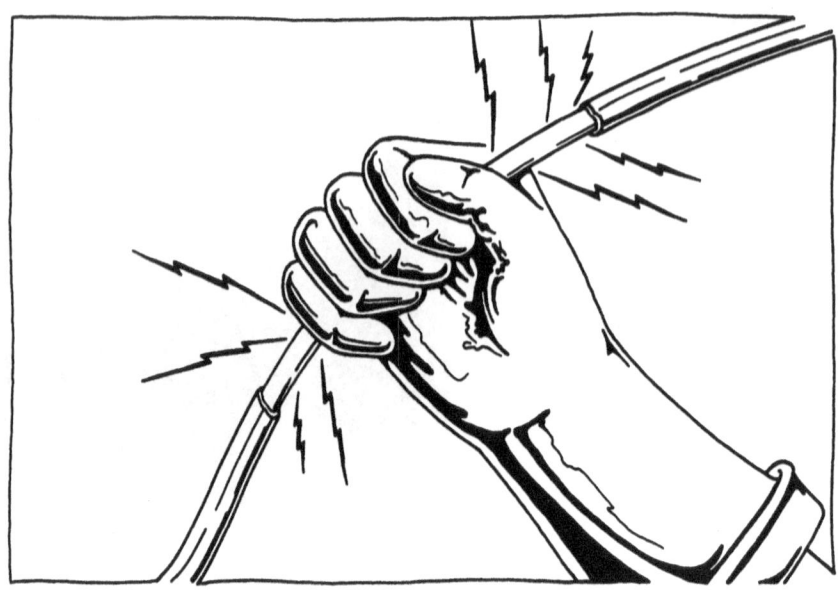

A: Because molecules of these substances hold their electrons tightly, so they have no free electrons to carry a current.

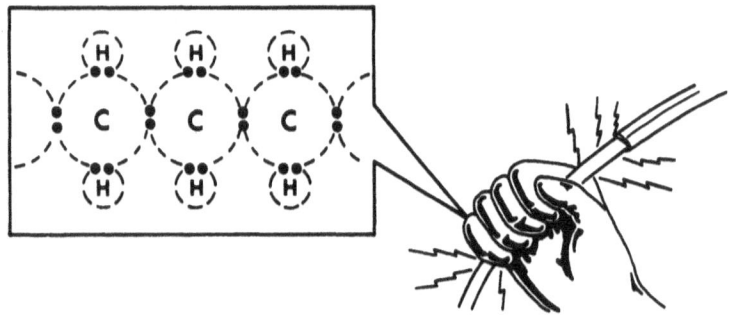

Hydrocarbon linkages, common in rubber and plastic.

Insulators have electron shells securely locked—unlike conductors, which are mostly metals with a few "stray" electrons in outer shells, far from the nucleus and easily detached.

Q: "Generating" electricity is somewhat a misnomer. The electrons already exist—generation merely induces them to move. How's it done?

A: Two ways, chiefly.

In a battery, chemical reaction with the acid adds electrons to the zinc plate, giving it a negative charge, and draws electrons from the copper, giving it a positive charge.

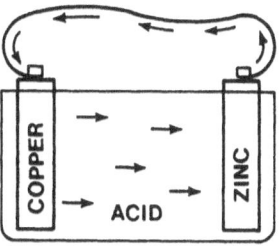

In a generator, a wire loop is rotated so it cuts through the lines of force between two magnets, which causes the free electrons in the wire to flow as a current.

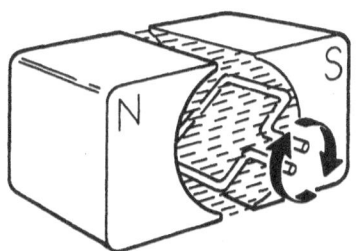

Q: What is lightning?

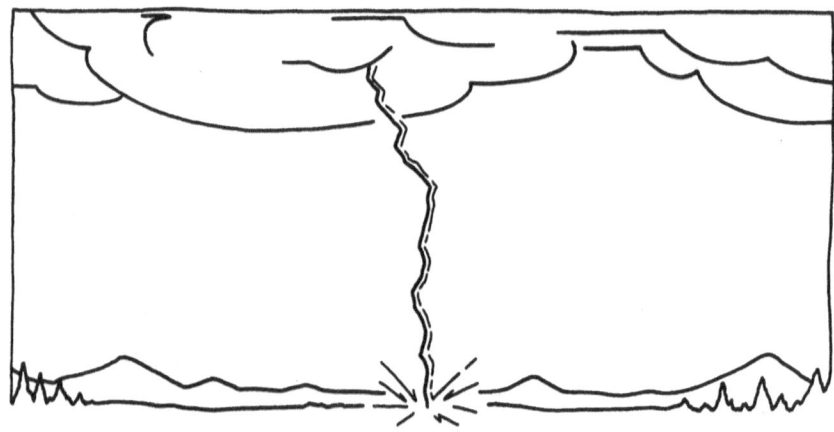

A: Giant leaps of electrons.

Theory: Clouds become charged because rising air shatters water droplets and ice crystals. The fragmentation leaves more electrons in the heavier particles, which settle in the cloud's bottom, charging it negative—which induces a positive field on Earth below.

First, electrons dart downward, ionizing a path through the air. Then a blinding "return stroke" leaps upward. As many as forty strokes may follow almost instantaneously until cloud and ground are neutralized. Air in the path, heated to 60,000°F, expands violently, making thunder.

Q: What is magnetism?

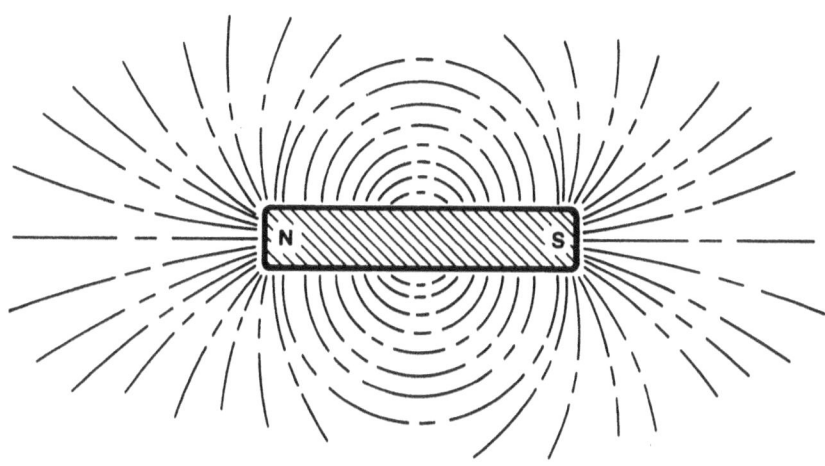

A: The power of electron "spin."

In atoms of most elements, electron pairs have opposite spin, balancing each other. But iron and a few other metals are unusual because some of their electrons spin in the same direction without compensation, giving each atom a magnetic field. A piece of iron becomes magnetized when a strong magnetic field causes its atoms to align their polarity in the same direction.

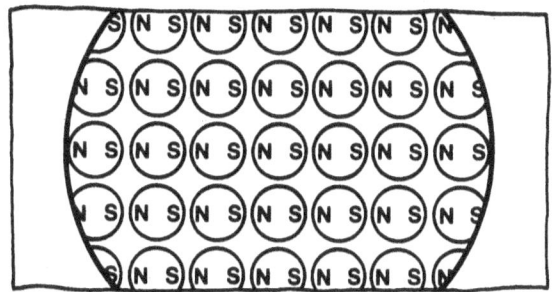

(Why is the whole Earth a magnet? The reason is unclear, but it isn't caused by this sort of atom-alignment.)

Q: How are magnetism and electricity mysterious partners?

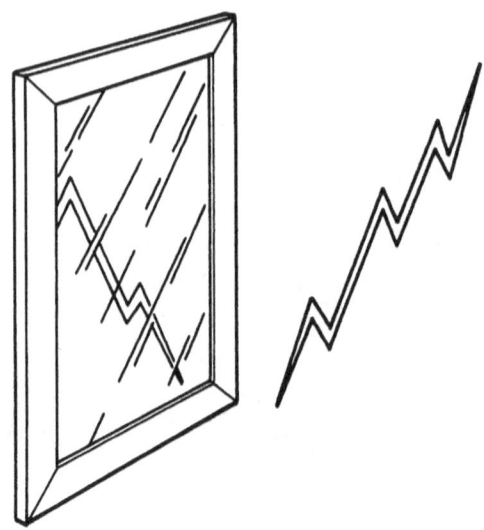

A: Each creates the other.
A moving magnetic field induces electron flow in a wire

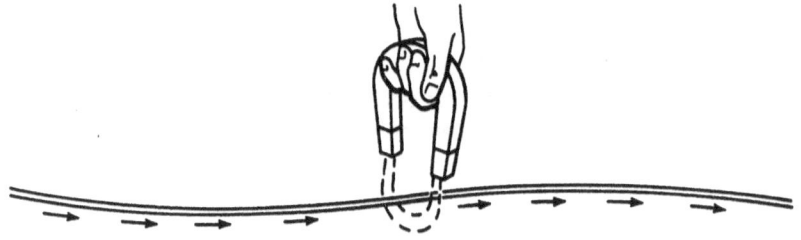

—and, conversely, flowing electrons in a wire create a magnetic field, which can be concentrated in an electromagnet.

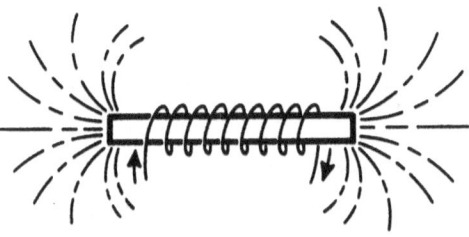

Both actions are functions of the electromagnetic force that creates light and governs electrons in atoms.

Q: What are the benefits of "superconductors," materials that offer no resistance to the flow of electricity?

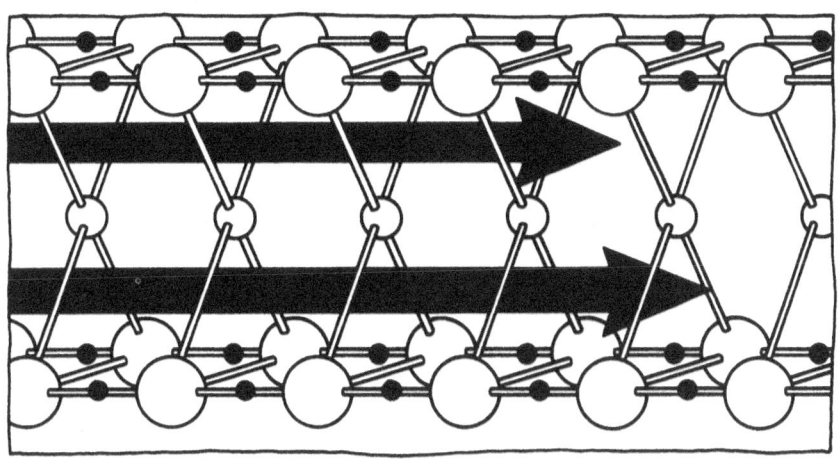

A: When a current is induced in a superconductor, it flows *forever*—even after the original power has been turned off.

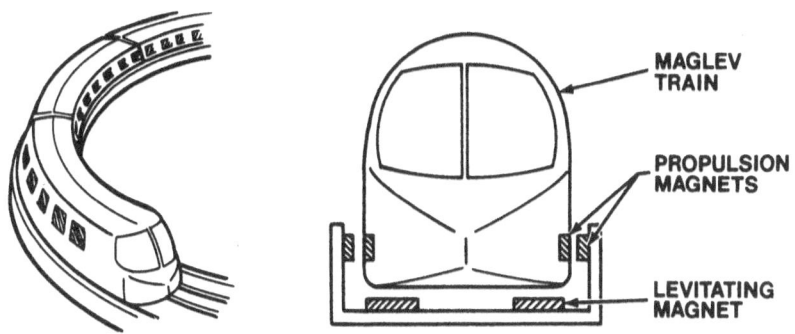

Thus superconductors can make powerful electromagnets needing little energy—for instance, the type for "maglev" trains in which magnetism levitates cars above rails.

It is *awesome* that electrons spinning in unison can lift a heavy train as much as twelve inches and hold it in midair.

Q: Albert Einstein said:
"There is no more commonplace statement than that the world in which we live is a four-dimensional space-time continuum."

True?

A: True. It's understandable when you get past the mysterious-sounding phrase.

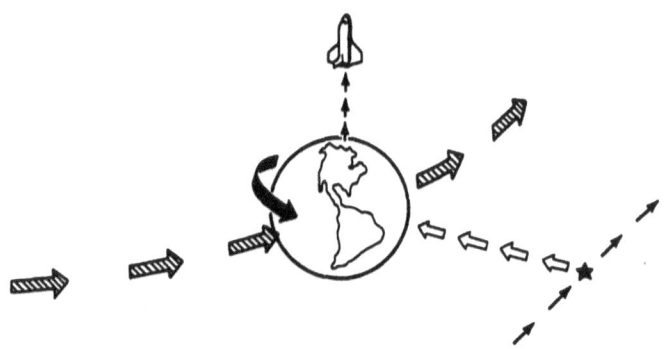

Physical things occupy three dimensions (length, width, and height)—but they're all in motion, and movement requires time, so the fourth dimension of time must be allowed for to comprehend correctly. This applies especially to the movement of light.

When you see a star, you see light that began traveling 10 or 10,000 years ago—and actually the star may no longer exist.

Q: Everyone knows that the world was transformed by Einstein's vision of matter being converted into energy.

How many more of his "impossible" concepts proved to be true?

A: Many.

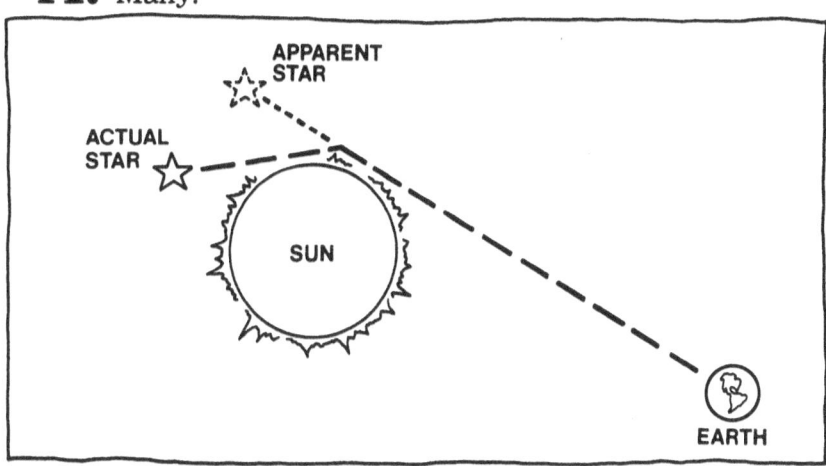

He said gravity bends light. People laughed—until sightings during eclipses proved it.

He said time slows and weight grows as speed increases. People laughed—until it was found that cyclotrons cause speeding particles to become heavier, with slower pulsations.

He said time runs faster away from Earth's gravity. People laughed—until tests with high-flying clocks proved it.

Q: Einstein failed in his lifelong quest for a "unified field theory" covering all the four forces in the universe.

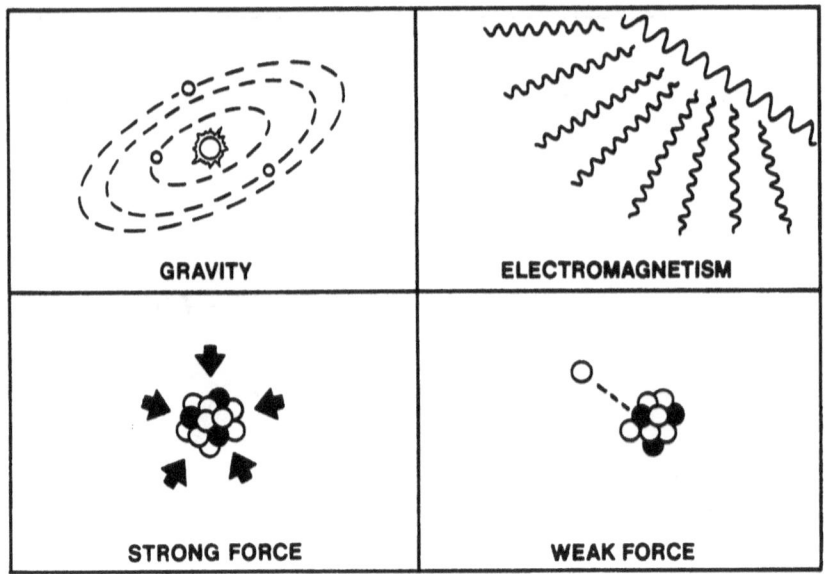

Has progress been made since his death?

A: Yes. It's becoming clear that each force is imparted by a *gauge particle* that is exchanged between matter particles.

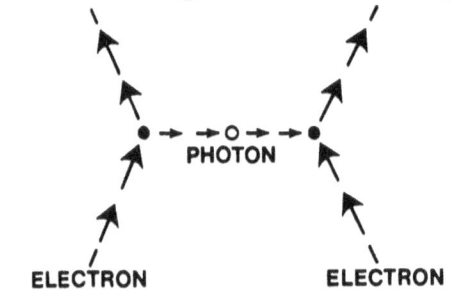

"Feynman diagram" of two speeding electrons exchanging a photon and being deflected.

Photons convey the *electromagnetic force* that produces light and rules electrons in atoms.

Gluons provide the mighty *strong force* that binds the atomic nucleus together.

W and Z particles bear the *weak force* that causes neutrons to decay in the heaviest atoms, producing radiation.

Gravitons (still undetected) presumably carry *gravity*.

Q: Electromagnetism stems from the power of electrons and protons to emit and absorb photons, massless units of light. How does this force affect *everything in the universe?*

A: It holds electrons in the atomic orbits that make matter seem solid.

It rules chemical reactions, performed by the outer shell of electrons.

It governs electricity, the displacement of electrons, and magnetism, the power of electrons spinning in unison.

It propagates light and other electromagnetic radiation, the photons that are emitted when electrons oscillate, decelerate, or jump to different shells.

Q: How did the discovery of quarks—fractional-charge components of the atomic nucleus—simplify comprehension of matter?

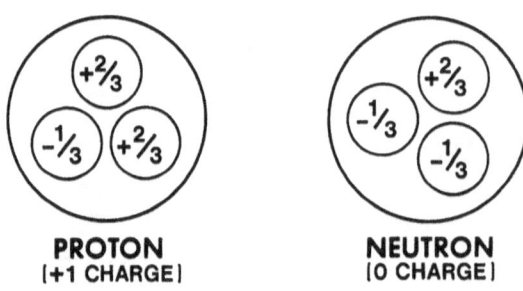

A: It revealed that virtually everything in our world is composed of just four matter particles:

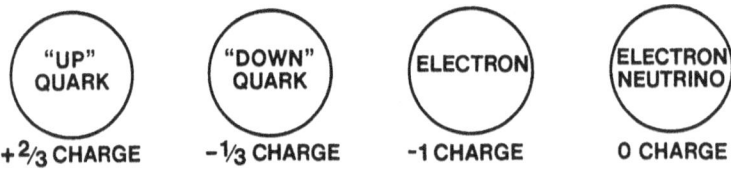

(Two other similar, heavier "families" of four particles each existed briefly after the big bang, but now they occur only in laboratory accelerators and cosmic rays. Also, each type of matter particle has an antimatter counterpart with opposite characteristics—but they likewise are rare in nature.)

Q: Microscopic cells—the fundamental units of all living things—are as complex as whole cities.

How can that be?

A: Most cells are so tiny that 500 would fit in the period at the end of this sentence—yet each contains about 200 billion molecules, making a multitude of metabolic machinery.

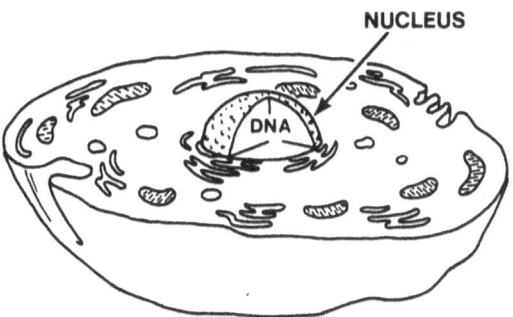

Cells are teeming factories that provide energy, manufacture proteins from amino acids, and perform other functions—then divide to make more cells.

The nucleus holds DNA (deoxyribonucleic acid), the key to all animals, all plants, all microbes, all life.

Q: DNA—the famous double-helix—is a threadlike molecule only a trillionth of an inch thick, but very long. Each human cell has about six feet of DNA tightly coiled in forty-six chromosomes. Since the human body has more than ten trillion cells, every person contains *several billion miles* of DNA.

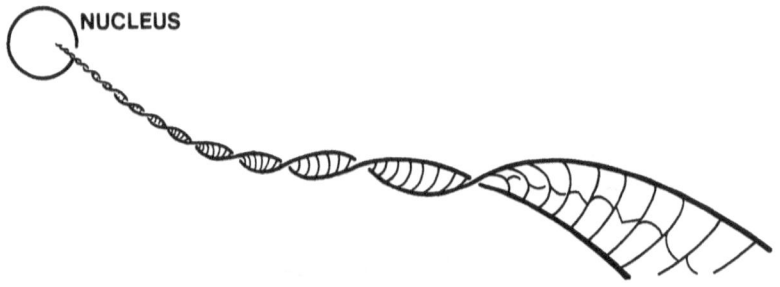

How does DNA convey the genetic codes that govern and shape every species?

A: When uncoiled, DNA molecules are like ladders. The rungs are links by four chemicals, two pairs which bind only with each other.

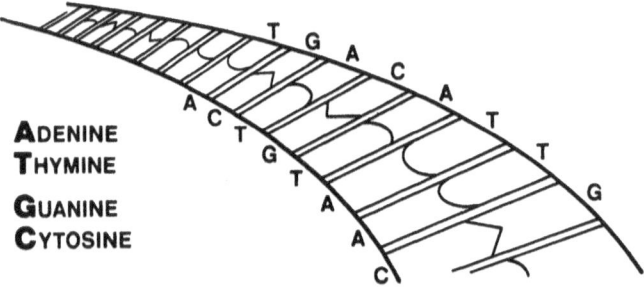

ADENINE
THYMINE
GUANINE
CYTOSINE

As a four-digit code, they spell out combinations of amino acid molecules needed for different proteins that compose the organism.

The code sequence for one protein is a "gene."

Q: How does the four-digit code copy itself and dictate the molecular sequences of proteins?

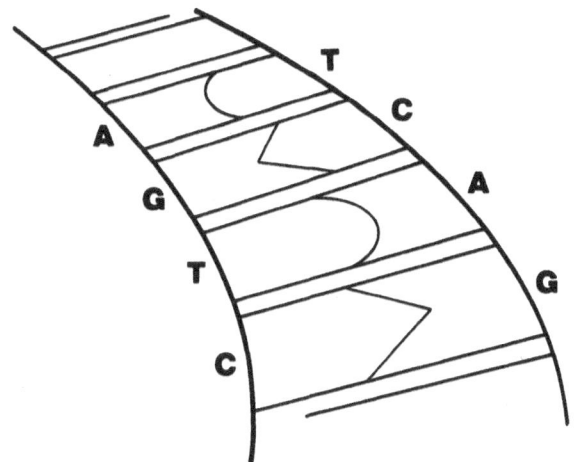

A: The ladder *unzips*, and new A, C, T or G molecules attach to appropriate spots, making an exact copy.

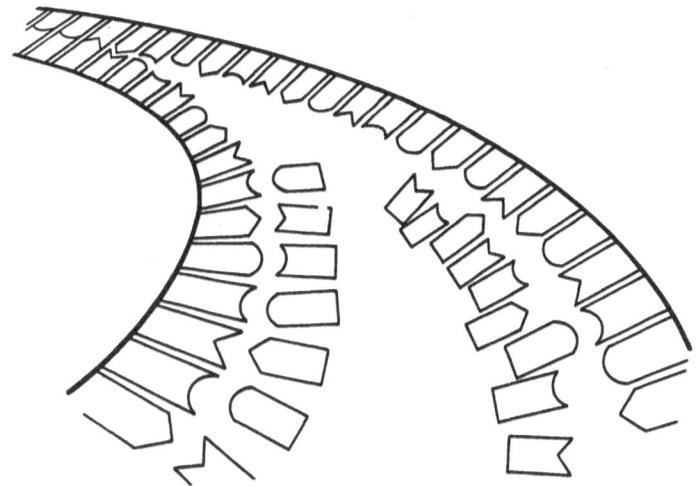

Just before a cell divides, the entire code is replicated for the new cell.

To manufacture a protein, a short segment (one gene) unzips and a temporary pattern is made, which is employed in selecting the amino acid molecules.

Q: How is this process manipulated in the booming field of genetic engineering?

"Supermouse" grew in 1983 from an embryo implanted with a human growth hormone gene.

A: Several ways. Here's an example of the chief method:

A human DNA molecule is subjected to an enzyme that cuts out the gene (sequence of signals) for manufacture of insulin. Other enzymes are used to insert the human insulin gene into DNA strands of E. coli intestinal bacteria.

In a laboratory vat, the E. coli makes millions of copies of itself bearing the foreign gene, and the E. coli cells dutifully assemble amino acid molecules into perfect human insulin, ready to extract and market.

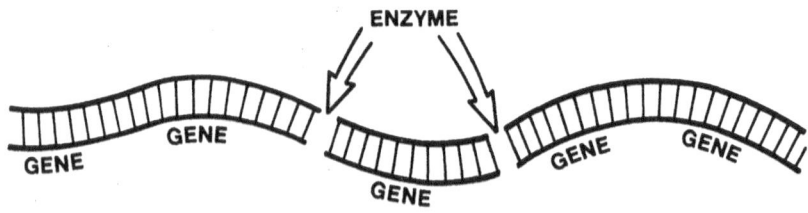

Q: Bacteria (one-celled plantlike organisms) cause pneumonia, tuberculosis, typhoid fever, diphtheria, whooping cough, syphilis, gonorrhea, leprosy, tetanus, etc. These diseases are treatable with antibiotics, which kill bacteria.

But virus-caused diseases—colds, flu, hepatitis, herpes, measles, chicken pox, smallpox, rabies, etc.—are harder to fight. How do viruses function?

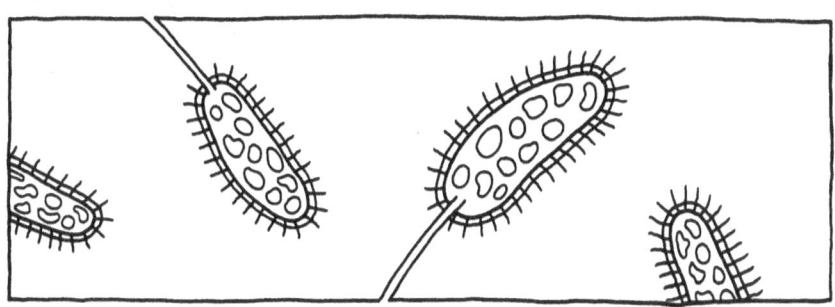

A: Viruses, one-tenth the size of bacteria, aren't quite alive. They are merely protein-covered packets of DNA (or related genetic molecules called RNA).

When a virus encounters a living cell, the viral DNA or RNA causes the cell's manufacturing process to make many new copies of the virus, which are emitted and infect other cells.

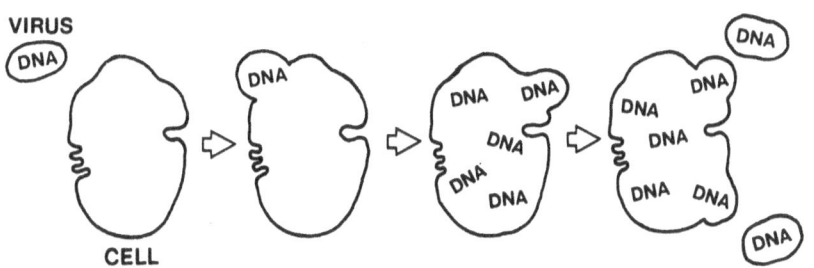

Q: What electric current creates the food of all living things?

A: In photosynthesis in plants, sunlight hitting chlorophyll molecules excites outer electrons, which jump to "acceptor" molecules and shed their excess energy as they travel through a mosaic of molecules.

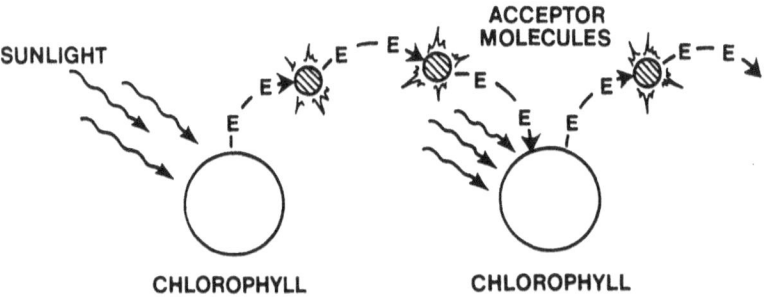

This flow of energy is channeled into plant processes that convert water (from the soil) and carbon dioxide (from the air) into sugar, starch, oxygen, and other compounds.

Plant sugar feeds not only plants but also the animal kingdom.

Q: What "power pack" carries the energy used in all animal and plant activity?

A: The "tail" of the ATP/ADP molecule.

Energy is poured into binding an extra phosphate group onto adenosine diphosphate, making it adenosine triphosphate. When cells need energy for their work, an enzyme severs the extra phosphate, loosing the stored energy.

Q: What electric current controls animal life?

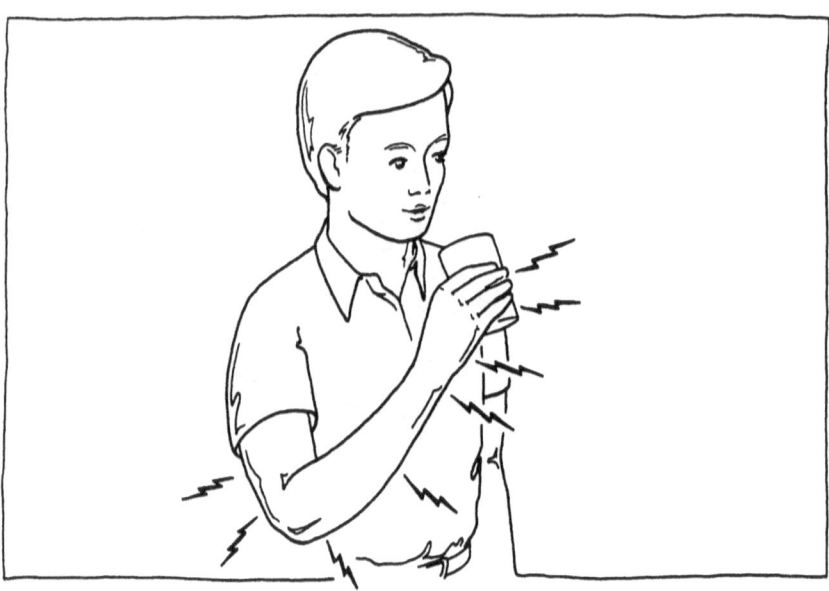

A: A signal flow of negative ions passed by long, long nerve cells controls every heartbeat, every breath, every movement in complex animals.

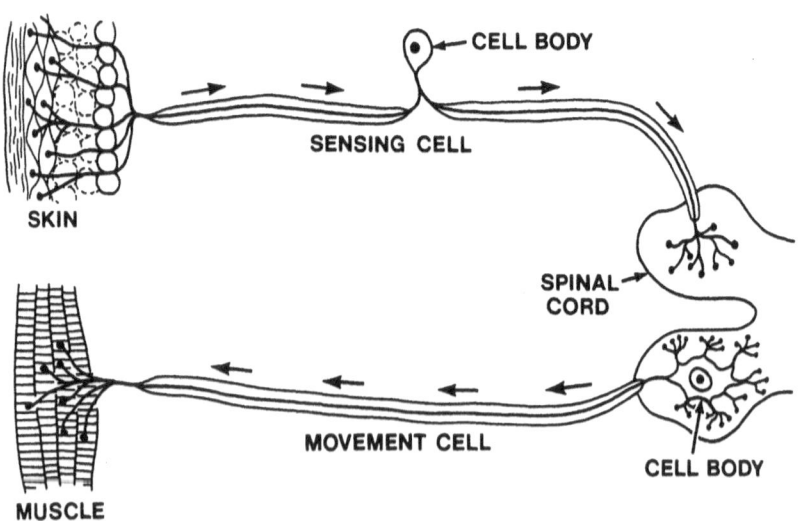

Q: Why do some scientists say a reptile brain and a mammal brain lurk inside the human mind?

A: During hundreds of millions of years of evolution, larger brain expansions surrounded previous stages, which continued functioning deep within.

The powerful human brain—with its ten billion neurons and 100 trillion circuit connections—includes buried components evidently inherited from reptiles and early mammals. The ancient parts seem to affect basics such as sex, fear, rage, survival, and the urge to nurture the young.

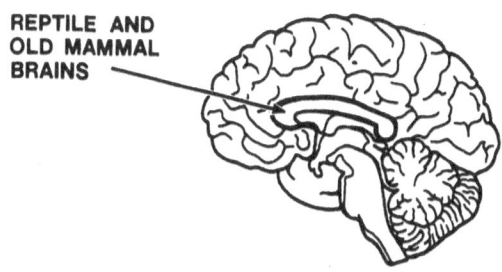

REPTILE AND OLD MAMMAL BRAINS

Q: How is sex causing the worst threat to Planet Earth?

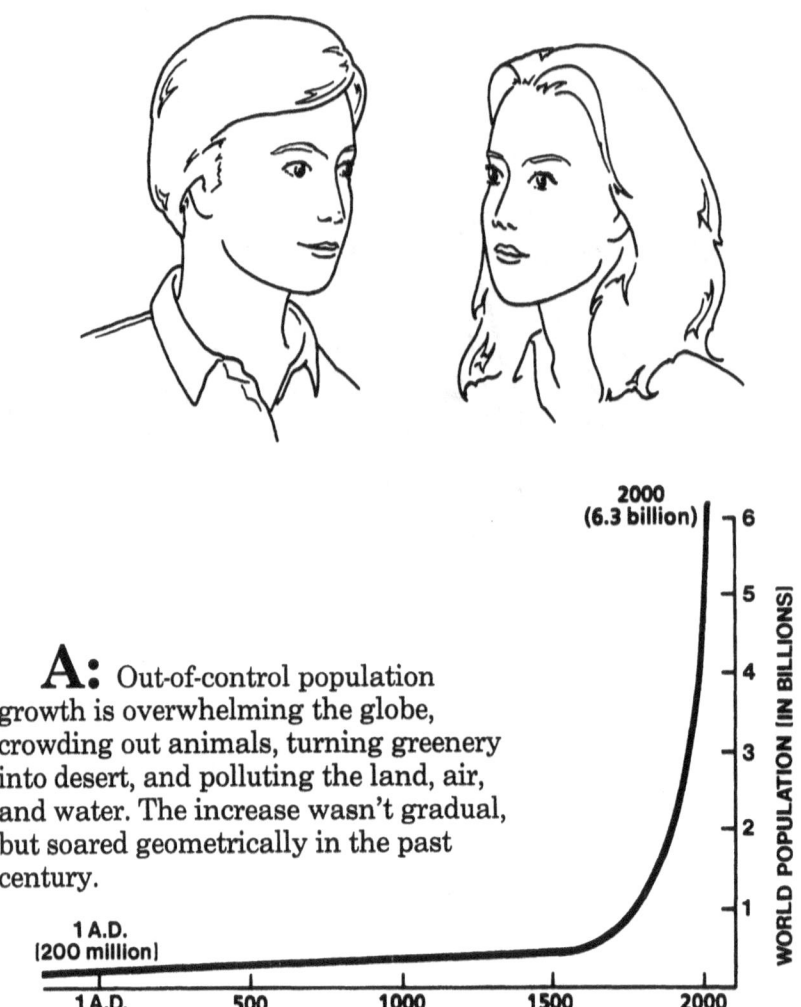

A: Out-of-control population growth is overwhelming the globe, crowding out animals, turning greenery into desert, and polluting the land, air, and water. The increase wasn't gradual, but soared geometrically in the past century.

Unless scientific methods of birth control are put to use everywhere—and soon—mass starvation and suffering are forecast in many poor nations.

Q: What is the greatest science accomplishment?

A: Perhaps this:

The average lifespan was just forty-eight years at the beginning of this century, a time when epidemics and childhood deaths still were rampant.

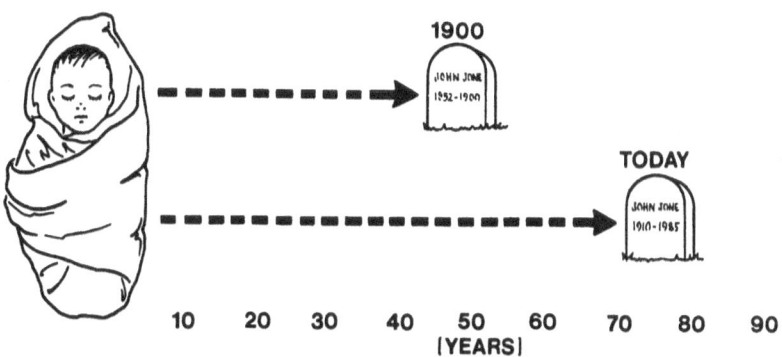

Today it's seventy-five and rising, thanks largely to the discovery of penicillin and other antibiotics that made it possible to control infections.

Q: Infinity lies in two directions—outward and inward.

What steps to infinity have unfolded, one after another?

A: From too large to imagine, to too small to imagine—and all dimensions in between—the role of science is to discover what exists and what mathematical laws govern.

In the bewildering complexity, there is order. Beneath the chaos are rules we can learn. Albert Einstein said: "The most incomprehensible thing about the world is that it is comprehensible."

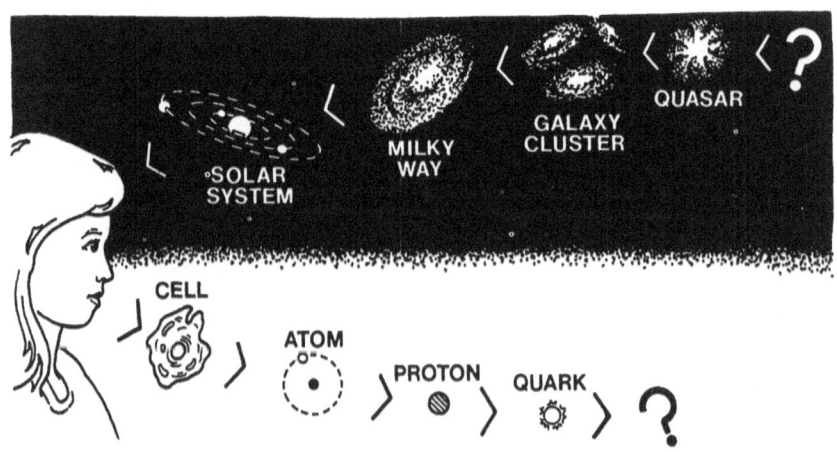

www.ingramcontent.com/pod-product-compliance
Lightning Source LLC
Chambersburg PA
CBHW031257290426
44109CB00012B/621